S0-AKG-393

Business Issues, Competition and Entrepreneurship Series

IMPROVING INTERNET ACCESS TO HELP SMALL BUSINESS COMPETE IN A GLOBAL ECONOMY

BUSINESS ISSUES, COMPETITION AND ENTREPRENEURSHIP

**Improving Internet Access to Help Small
Business Compete in a Global Economy**
Hermann E. Walker (Editor)
2009. ISBN: 978-1-60692-515-7

Business Issues, Competition and Entrepreneurship Series

IMPROVING INTERNET ACCESS TO HELP SMALL BUSINESS COMPETE IN A GLOBAL ECONOMY

HERMANN E. WALKER
EDITOR

Nova Science Publishers, Inc.
New York

LIBRARY OF CONGRESS CATALOGING-IN-PUBLICATION DATA
Available upon request
ISBN: 978-1-60692-515-7

Published by Nova Science Publishers, Inc. ✦ New York

CONTENTS

PREFACE

Businesses are quickly integrating new telecom services into their business plans. Broadband connects entrepreneurs to millions to distant customers, facilitates telecommuting, and increases productivity in many ways. Much of our economic growth is attributable to productivity increases arising from telecommunications advances.

This new book presents the edited text of a U.S. Senate Hearing devoted to this subject - *Hearing Before he Committee On Small Business And Entrepreneurship United States Senate One Hundred Tenth Congress, First Session, September 26, 2007*

OPENING STATEMENTS

Wednesday, September 26, 2007
United States Senate,
Committee On Small Business And Entrepreneurship,
Washington, DC.

The Committee met, pursuant to notice, at 10 a.m., in room 428– A, Russell Senate Office Building, the Honorable John F. Kerry (Chairman of the Committee) presiding.

Present: Senators Kerry, Snowe, and Corker.

OPENING STATEMENT OF THE HONORABLE JOHN F. KERRY, CHAIRMAN, SENATE COMMITTEE ON SMALL BUSINESS AND ENTREPRENEURSHIP, AND A UNITED STATES SENATOR FROM MASSACHUSETTS

Chairman KERRY. The hearing will come to order. Thank you all very much for joining us this morning to discuss how we are going to improve Internet access for small businesses in the country and the importance of being able to be online for business in today's world.

I would just direct you quickly to today's New York Times and the Business Section, "Strategies to Succeed Online." In the middle of the article it says the old ways of hiring a public relations firm and putting out press releases just don't cut it anymore. Today's businesses have to be more hands-on, grassroots, interactive, and maintain this flow of continuous communications.

That is what this hearing is all about.

Today, the Committee is exploring the pivotal, critical question of access for small businesses to the Internet. We want to look at the question of whether the prices are affordable, to what degree there is penetration, are the speeds adequate, and what do we do in order to make improvements?

Most people don't disagree that high-speed Internet access is critical to economic competitiveness. You hear it talked about all the time and everybody in public life has it in their speeches. But they don't necessarily have it in their policies, and for small business, increasingly, it is becoming critical in order to track inventory, create jobs, monitor consumer relations, forecast product sales— any number of different things. The Internet is not a luxury, it is a necessity. It is imperative in maintaining our growing economy In March of 2004, President Bush appeared to understand that by setting forth the Universal Broadband Access Goal by 2007. Well, we are in 2007, but we have yet to put in place the policies that will actually realize that goal. So as a result, we are lagging behind the rest of the world now, which is pretty incredible when you consider I remember sitting in the Commerce Committee in 1996 when we wrote the Telecommunications Act, mostly thinking about telephony; within months, it was blown away and almost obsolete because it was all data transformation and data transmittal that really was at stake. And here we are now, just a little more than 10 years later, and the United States is lagging behind.

When the Organization for Economic Cooperation and Development, the OECD, began surveying and ranking broadband use, the United States was ranked fourth among the 30 nations surveyed, behind Korea, Sweden, and Canada. Since 2000, the United States has plummeted in the OECD rankings to 15th place, and another ranking of access to high technology lists the United States 21st, behind Estonia and tied with Slovenia.

We can do better than this and we have to do better than this. It is almost shameful, folks. It is inexplicable. It is essential for America to have a national broadband strategy that encourages competition and expands broadband access, or we are going to continue to be left behind.

Today, from rural areas to big cities, nearly 60 percent of the country does not subscribe to broadband service, in part because they simply don't have access to the service or they can't afford it. Even a nationwide leader in technological innovation like my home State of Massachusetts had a 45.9 percent broadband penetration rate at the beginning of 2006, and that was the fourth best rate in the country.

While small businesses are the backbone of our growing economy, the power of the tools that they use to compete both domestically and globally are shrinking dramatically. With America's Internet speeds severely lagging behind universal standards, it is surprising that small businesses can compete at all. Americans in rural communities face especially difficult challenges in overcoming problems with broadband deployment, since many lack even basic access.

The outcome is clear. We place a technological ceiling on job growth, innovation, and economic production. We cannot expect small businesses to fairly compete against more technologically advanced competitors unless we change what is happening today.

Some experts estimate that universal broadband would add $500 billion to the U.S. economy and create 1.2 million jobs. With numbers like those beckoning us, we need to focus on reestablishing our technological edge.

I am delighted that we have two FCC Commissioners here today on the first panel, Michael Copps and Jonathan Adelstein, to tell us what they feel needs to be done to develop a national broadband strategy. And on our second panel, we are pleased to welcome Ben Scott, who is a recognized leader in broadband deployment and media issues; Doug Levin, the CEO of Black Duck Software, who will give us a unique perspective as a technology business leader; Mr. Mefford will talk about innovative approaches to broadband being pursued in Kentucky; and Mr. Wallsten with the Progress and Freedom Foundation offers additional ideas on the current state of Internet penetration. We look forward to hearing their testimony.

A few things are certain here. We need better information in the development of these policies. We are broadly lacking broadband data for small business itself. I plan to ask the Small Business Administration and the FCC to conduct a robust effort to gather data about small business and broadband usage.

We also need a strong regulatory framework to encourage competition. Competition spurs innovation, enhances services andd reduces prices. I have advanced and supported a series of measures designed to increase competition. For example, I have worked to make better use of spectrum, which is a valuable public asset. Much of our spectrum is underutilized, shelved, and hoarded by in-cumbent companies. We can maximize this valuable asset, including the use of the white spaces, by creating 700 megahertz auction rules that encourage new market entrants; in fact, we are dealing with some of that on the Commerce Committee.

Lastly, we need to think creatively about Internet access. We ought to look at reforms of the Universal Service Program and innovative public-private partnerships for additional ideas. I hope we can draw these and other issues out in the hearing. I look forward to hearing from our witnesses.

Senator Snowe, good morning and thanks for being with us.

Opening Statement of the Honorable Olympia J. Snowe, a United States Senator from Maine

Senator SNOWE. Good morning. Thank you, Mr. Chairman, for holding a hearing on this vital issue. I hope our combined membership on both the Commerce and Small Business Committee can help us work together and develop a policy with respect to broadband deployment.

I want to thank Commissioners Copps and Adelstein from the FCC for their tremendous stewardship and public service. I have the highest respect for both of these Commissioners and I want to thank them for recently holding a hearing in Portland, Maine, to solicit testimony from various segments of the population regarding key telecommunications issues and preserving localism in the media marketplace.

I have known Commissioner Copps for some time now and I applaud his unwavering leadership on the Federal-State Joint Board on Universal Service and in particular the E-rate program and his efforts to expand the Universal Service Fund to include broadband deployment. Commissioner Adelstein's understanding and experience with rural broadband deployment is highly essential and key voice in the FCC and I want to thank you, as well, for your steadfast dedication and commitment to expanding broadband across America.

I look forward to a productive and constructive dialogue with the Commissioners and other expert witnesses on ways in which the Federal Government can encourage more robust broadband deployment, specifically to rural America and businesses The President announced his priority 3 years ago for broadband deployment by 2007. We have a goal, but not the tactics to realize this initiative. Fulfilling this charge is imperative as small businesses who rely on broadband connections, specifically in rural areas such as Maine, need affordable access to technologies of the future and, as well as the ability to compete in the global marketplace where other countries and our international counterparts have a national broadband strategy.

One of the issues associated with universal broadband deployment is, of course, the FCC's lack of a comprehensive broadband data gathering methodology. I know both Commissioners have been an advocate of making improvements in this area. The GAO agreed in November of 2006, indicating that without more reliable data, the FCC is unable to determine whether its regulatory policies are achieving their goals.

I would like to explore the FCC's adherence to the Regulatory Flexibility Act, which requires the Federal agencies to consider the effect of these proposals on small businesses. Commissioners Copps and Adelstein, you are at the forefront of these issues and I welcome your input on how small businesses can work with the FCC to reap the benefits of broadband services.

As Ranking Member of this Committee, I firmly believe that Federal policy should promote a universal broadband market that deploys competitive and affordable broadband. Today, the marketplace lacks competition, with 98 percent of Americans receiving their broadband service either from a cable or phone company. To encourage growth, we need to promote more competition in the market.

I am particularly pleased that many States and municipalities have launched initiatives to bring high-speed Internet services and economic opportunity to communities the market has overlooked. One example of this growing trend is Connect Maine, an ambitious public-private partnership which seeks to provide 90 percent of Maine's residents with broadband access by 2010.

As we consider the matter of competitiveness, we must also bear in mind that affordability is as much a barrier. According to a report by the Small Business Administration's Office of Advocacy, rural small businesses do not subscribe to broadband services as frequently as urban small businesses do, usually because of the high cost, creating a digital divide. In Maine, for example, even in the areas where they do have access to broadband, 59 percent choose not to subscribe because of the high cost. So, we must work together to address the disparities between those who have this access and those who do not.

As many will mention here today, the United States, ranks very poorly in broadband penetration, although it raries by ranking the International Telecommunications Union ranks the United States 15th in terms of global broadband penetration rate. That is an unacceptable ranking, in the 21st century, for the United States globally to be ranked 15th in a category where it has been a pioneer.

In Maine, the statistics are just as bleak. It ranks 31st in the country for residential broadband penetration, and 14 percent of households have no access whatsoever. In America, it is 1 in 10 consumers who have no access So, as we can see, broadband deployment in Maine and throughout the country is severely lacking. It continues to be one of the major concerns among small businesses in my State, and rightfully so, because broadband investments can have a substantial economic impact.

Everybody agrees that broadband holds the promise of technological innovations, better communication, and connecting vast distances within the States. So the question for this Committee is how do we engender and promote a robust market, create that policy that charts a path to successfully deploying broadband to under- served small businesses?

Hopefully, this is just the beginning of this dialogue and we can chart this policy. I think it is absolutely crucial that we begin the process in a very efficient and expeditious way, and hopefully it can be spurred by this Committee hearing this morning.

Mr. Chairman, thank you.

Chairman KERRY. Thank you, Senator Snowe.

I would like to try to go right to the witnesses. Are you amenable?

Senator CORKER. Yes.

Chairman KERRY. Great. Gentlemen, thank you for being with us. We look forward to your testimony. Your full statement will be placed in the record, as if read in full. If you could summarize in about 5 minutes, we would appreciate it.

Commissioner Copps.

TESTIMONY

STATEMENT OF THE
HONORABLE MICHAEL J. COPPS,
COMMISSIONER, FEDERAL COMMUNICATIONS
COMMISSION, WASHINGTON, DC

Commissioner COPPS. Chairman Kerry, Ranking Member Snowe, Senator Corker, thank you for holding this hearing. Time is short, so I will be blunt.

America's lack of a coordinated broadband strategy is imposing huge costs on small businesses all across the land. As the front page of the Washington Post recently stated, "Americans invented the Internet, but the Japanese are running away with it." The most recent broadband rankings by a variety of organizations have the United States at anywhere from 11th all the way to 25th, and all of them have us falling. This is not where your country and mine is supposed to be.

It is not just a matter of national pride that we are talking about, it is a business issue. Small businesses everywhere are increasingly relying on broadband Internet access. It is as essential as running water, electricity, or phone service. Some small businesses in rural America cannot get an Internet connection at all, and even when they can, they typically pay too much for service that is too slow. It isn't that much better in the Nation's metropolitan areas. Prices are high for service that is by global standards uncompetitive.

The Internet is supposed to be our great equalizer, leveling the playing field between urban and rural, large and small, and domestic and global businesses. The broadband system we have today makes a mockery of this great promise by creating greater disparity How do we turn things around? We need a comprehensive national strategy and a strong commitment from the very top that broadband is our national infrastructure priority. We need all the departments of Government cooperating to encourage broadband deployment using whatever mix of grants and incentives Congress may choose.

There is an important role for the FCC. The Commission owes Congress and the country more than they are getting. First, better data. The Commission still unbelievably defines broadband as 200 kilobits per second. How 1997 that sounds. The Commission still assumes that if one person in a ZIP code has broadband, ergo, everybody has it. So let us get better definitions of speed and deployment and granular data on prices, and let us study also what other nations are doing, because there are some lessons to be learned there.

Second, the FCC needs to become a clearinghouse for all the broadband innovation and experimentation that are occurring outside the beltway. I have attended broadband summits and met with local experts and small business owners in Cambridge, Massachusetts; Portland, Maine, and all around the country. I have learned that our diverse and varied Nation has immense reserves of local creativity. It is time to start sharing and encouraging that creativity.

Third, the FCC needs to bring competition back into its telecom policies. For example, the GAO has demonstrated that the FCC's deregulatory policies and our approval of one big merger after another have saddled small businesses with increased costs, like special access prices. The Commission is scheduled to act on special access soon, and I hope Commissioner Adelstein and I can find a majority willing to stand up for entrepreneurs and consumers, not just incumbent phone companies.

Fourth, we need to support broadband with the Universal Service Fund. It worked for plain old telephone service, and it will work here. I am delighted that the Federal-State Joint Board recently agreed with me that broadband must be the mission of the USF for the 21st century. We need to make that happen soon. Congress gave the FCC considerable authority to get broadband out to our people, and we need to start using that authority aggressively.

You know, throughout our Nation's history, we have always found ways in this country to work together, business and government and communities, to build our physical infrastructure, whether it were roads or turnpikes or canals way back when, as well as railroads, and highways. Why can't we tackle this infrastructure challenge the same way, pulling together to get the job done instead of assuming that it is somehow just going to magically happen all by itself. It is not happening, and it needs to.

I want to mention one more issue, not in my prepared statement, but I talk about it wherever I go, and it has real small business implications. It appears that the FCC may be asked to vote on media ownership issues soon, perhaps by the end of the year. Last time we did that, in 2003, it was a disaster from which we were rescued by the Senate and the courts. Media is not just another industry, it is the most potent social, political, and cultural influence in the country. It is how we communicate, inform, debate, and de cide. Arthur Miller once said that a good newspaper is a nation talking to itself, and that is really what media is.

Increasingly, media has become the province of a few mighty conglomerates who have sacrificed much of the localism and diversity and small business competition that are supposed to be the bedrock of our TV and radio, and the FCC has aided and abetted that at every step of the way. This has been nothing short of a disaster, not only for small businesses, but for our culture as a whole. The rise of big media has encouraged the homogenization of local journalism, arts, and culture and led to the degeneration of America's civic dialogue.

It has been a special disaster for minority businesses. People of color are 30 percent of our country's population, but they own 3.26 percent of all full-power commercial television stations. Is it any wonder that TV is so full of caricatures and distortions?

As you, Mr. Chairman, and Senator Obama pointed out in a letter to us, the FCC has had an open proceeding for years on how to increase media ownership by small businesses, women, and minorities. You called upon the FCC to complete this proceeding and make headway on the appalling situation we face today before we make further changes to our rules. I support your call 100 percent. I know my colleague, Jonathan Adelstein, feels strongly about this. It is time to draw a line in the sand, be honest about what is at stake, and not proceed on media ownership until we figure out how to get a seat at the table for women, minorities, and small businesses.

My time is up, but I did want to get on the record that whether it is broadband or broadcast, small businesses are up against challenges not of their own making, and they are suffering and suffering badly as a result. We can do better. We must do better. Thank you.

[The prepared statement of Commissioner Copps follows]

TESTIMONY OF FCC COMMISSIONER MICHAEL J. COPPS

TESTIMONY OF FCC COMMISSIONER MICHAEL J. COPPS

**U.S. SENATE COMMITTEE ON SMALL BUSINESS AND
ENTREPRENEURSHIP**

**IMPROVING INTERNET ACCESS TO HELP SMALL
BUSINESS COMPETE IN A GLOBAL ECONOMY**

SEPTEMBER 26, 2007

Chairman Kerry, Ranking Member Snowe, and Members of the Committee, thank you for inviting me here to talk about broadband and America's small businesses. I'll be brief and blunt. America's lack of a broadband strategy is imposing huge costs on small businesses all across the land.

As the front page of the *Washington Post* recently stated, "Americans invented the Internet, but the Japanese are running away with it.... Accelerating broadband speed in [Asia and much of Europe] ... is pushing open doors to Internet innovation that are likely to remain closed for years to come in much of the United States."[1] Indeed, the most recent broadband rankings by international organizations, think tanks, industry groups and business analysts have us at 11th, 12th, 15th, 20th, 24th, and 25th in the world.[2] Take your pick of studies, but this is not where the United States is supposed to be. And the trend lines of these studies are plunging downward.

This isn't just a matter of national pride. It's a business proposition, a competitiveness issue. Our lackluster broadband performance is a huge barrier to, and tax upon, innovation and entrepreneurship. Businesses everywhere are increasingly reliant on broadband Internet access; it has become as essential as electricity, running water or phone service. Yet many small businesses in rural America cannot get an Internet connection at all. Even where they can, they typically pay too much for service that is too slow. The story isn't all that much better in the nation's metropolitan areas. Prices are high for service that is, by international standards, uncompetitive.

The Internet *should* be the great equalizer—leveling the playing field between urban and rural; large and small; domestic and global businesses. The broadband system

---

[1] Blaine Harden, "Japan's Warp-Speed Ride to Internet Future," *Washington Post* at A1 (August 29, 2007); *see also* Jessica E. Vascellaro, "Is High-Speed Internet Growth Slowing," *Wall Street Journal* at B3 (August 9, 2007) ("Industry watchers predict broadband growth will continue but statistics indicate that the U.S. will remain well behind other countries that adopted broadband more quickly. The U.S. is ranked 25th in broadband penetration, behind countries including South Korea, where penetration is 89%, and Canada, where it is 63%.").

[2] FTTH Council, "Asia Lead the World in FTTH Penetration," (July 18, 2007) (ranked 11th); Robert Atkinson, "The Case for a National Broadband Policy," (June 2007) (ranked 12th); OECD, "Broadband Statistics to December 2006" & ITU, "Broadband Statistics for 1 January 2006" (ranked 15th); ITU United Nations Conf. on Trade and Develop., "Chapter 3, the Digital Opportunity Index," *World Information Society 2007 Report: Beyond WSIS*, p. 36. (ranked 20th); Website Optimization, LLC, "US Jumps to 24th in Worldwide Broadband Penetration" (August 21, 2007) (ranked 24th); *supra* n.1 (ranked 25th).

we have today makes a mockery of this great promise and instead creates competitive disparities.

Part of our problem is reliance upon duopoly and oligopoly where we should be enjoying vigorous carrier and network competition. As part of the recent 700 MHz auction, the FCC heard arresting testimony from a wireless entrepreneur who explained that the U.S. is way behind Europe when it comes to developing and marketing innovative wireless broadband devices.[3] The big losers are small companies squeezed out by the behemoths that have come to dominate the industry.

Several months ago, I visited Portland, Maine, and heard about a local stay-at-home mom who had developed a small retail business over the Web. Her market is limited to Americans who have high-quality Web access. Greater broadband penetration would help her, as well as millions of other entrepreneurs who lack bricks and mortar stores. Broadband is this era's bricks and mortar. One recent study concludes that every percentage point increase in broadband penetration (currently around 50% in the U.S.) would mean 300,000 more jobs and increased national output.[4]

How do we turn things around? Let's start with a comprehensive national strategy. We need a strong statement, combined with serious commitment from the very top—not just a campaign promise—that broadband is a national priority. We need to make sure all the branches of government are cooperating to encourage broadband deployment, using financial tools such as matching grants and tax incentives.

There are also a series of specific steps the FCC can take. The Commission owes you more than you are getting. First is improved data-gathering. Our current efforts are woefully out-of-date and out-of-whack. The Commission is still calling 200 kilobits per second "broadband" and assuming that if one person in a ZIP code has broadband access, ergo, everyone else does as well. This is 2007, not 1997. We need a more credible definition of speed and more granular measures of deployment, as well as to start gathering data on price and the experience of other nations. There's a lot to learn there.

Second we need to start cataloging and benefiting from all the innovation and experimentation that's occurring outside of Washington D.C. Over the past year, I've attended broadband summits and met with local experts and small business owners in Cambridge, Massachusetts; Portland, Maine; rural Hawaii; Lawrence, Kansas; Little Rock; New Orleans; and Seattle. I've learned we live in a diverse and varied nation with immense reserves of local creativity. In some areas, experience indicates wireless broadband may be the answer; in others, it may be increasing competition among fiber

---

[3] Testimony of Jason Devitt at Federal Communications Commission July 31, 2007 Open Meeting, available at http://www.fcc.gov/realaudio/mt073107.ram at 9 minutes 30 seconds.

[4] Robert Crandall, William Lehr and Robert Litan, *The Effects of Broadband Deployment on Output and Employment: A Cross-sectional Analysis of U.S. Data* (Brookings Institution: July 2007), available at http://www3.brookings.edu/views/papers/crandall/200706litan.pdf.

providers. The FCC—with its 2,000 communications experts—ought to be playing a leading role as a clearinghouse for broadband ideas that have worked.

Third, there is enormous room to improve our competitive telecommunications policies. The GAO's examination of the special access market (for bulk telephone and broadband services) reveals that around 94% of commercial buildings are served exclusively by the incumbent telephone company.[5] The same report also demonstrates that the FCC's deregulatory policies, and its approval of merger after merger, have saddled small and medium-sized businesses with increased special access prices. The FCC is currently considering action in this area and I hope that Commissioner Adelstein and I can find a majority willing to stand up for small and medium-sized businesses—and ultimately for American consumers—rather than for incumbent telephone companies.

Fourth, we need to commit to supporting broadband with the Universal Service Fund. It worked for plain old telephone service and it will work here. I am delighted that the Federal-State Joint Board recently agreed with me on a bipartisan basis that broadband must be the mission of the USF for the 21st Century. I look forward to working with all my colleagues at the Commission to make this a reality.

Throughout our history, we have always in this country found ways to build our physical infrastructures: roads, turnpikes, canals, harbors, railroads, highways. Why can't we tackle this one the same way, with business, government and communities pulling together to get the job done?

---

[5] United States General Accountability Office, *FCC Needs to Improve Its Ability to Monitor and Determine the Extent of Competition in Dedicated Access Services* (GAO-07-80, November 2006), available at http://www.gao.gov/new.items/d0780.pdf.

Chairman KERRY. Thank you very much. I appreciate the direct and important testimony that you just gave.

Commissioner. Adelstein.

STATEMENT OF THE HONORABLE JONATHAN STEVEN ADELSTEIN, COMMISSIONER, FEDERAL COMMUNICATIONS COMMISSION, WASHINGTON, DC

Commissioner ADELSTEIN. Thank you, Mr. Chairman, Senator Snowe, and Senator Corker. Thanks for inviting me.

Mr. Chairman and Senator Snowe, I have certainly long admired your leadership on technology issues. You well understand that broadband is one of the best tools for promoting economic growth that we have ever seen in this country. It is a key factor in the success of so many of our small businesses.

Small businesses drive job creation, economic development, and new technologies, as hearing after hearing has demonstrated. They also purchase a massive amount of telecommunications services, $25 billion a year. So I am deeply concerned about the problems with prices, speeds, and availability of broadband services.

Unfortunately, as the GAO recently noted, the FCC collects very little reliable data about the availability of broadband to small business. We can't fix what we don't understand.

The good news is that businesses are quickly integrating new telecom services into their business plans. Broadband connects entrepreneurs to millions to distant customers, facilitates telecom- muting, and increases productivity in so many ways. As we know, much of our economic growth is attributable to productivity increases arising from telecommunications advances.

Given that 52 percent of our small businesses are homebased, broadband capability is critical. Just as the Pilgrims used the Mayflower to reach new opportunities in Plymouth Harbor, entrepreneurs are using broadband to reach beyond their current horizons.

Now, the bad news is that the little data we have suggests that small businesses are starved for telecommunications competition. Many small businesses have only one choice of broadband provider. This deprives them of innovative alternatives and can force them to pay higher prices. Even where there are competitive options, alternative providers rely heavily on inputs from incumbents, highlighting the importance of pro-competitive policies, as we have in the Telecommunications Act.

Our businesses now compete on a global stage, so we have got to tap the potential of all their citizens, no matter where they live. We need to prevent the outsourcing of jobs overseas by promoting the insourcing of jobs here within our own borders. While we have made some progress, I am very concerned that we are failing to keep pace with our global competitors, as you noted. Every year, we slip further down the international rankings. The bottom line is, citizens of other countries are simply getting more megabits for less money.

I am concerned that lack of a broadband plan is one reason we are falling behind. We need a comprehensive national broadband strategy, and to lay out some elements of it, it should incorporate benchmarks, deployment time tables, and measurable thresholds togauge progress. We need to set ambitious goals, magnitudes higher than the 200 kilobits we now count as broadband. We should gather better data, including better mapping of broadband availability, as you have up there for Massachusetts. We don't have good data for much of the rest of the country that was done by the private sector. The Government has little idea where broadband is truly available.

The FCC should be able to give Congress and consumers a clearer sense of the price per megabit, just as we look to the price of a gallon of gas as an indicator of consumer welfare. We must also increase incentives to invest, because the private sector will drive deployment. And we must promote competition, which is the best way to foster innovation and lower prices.

We must also ensure that universal service evolves to support broadband so that our hardest-to-serve areas are covered. As you noted, Mr. Chairman, spectrum-based services offer some of the best opportunities for promoting broadband. We must get broadband spectrum into the hands of operators ready to serve at the local level, including small businesses. One way is through auctioning smaller license areas that are affordable to community- based providers.

With the upcoming massive 700 megahertz auction, we have an historic opportunity to facilitate the emergence of a third broadband platform. I hope that companies will look at the rules that we made and we developed as a compromise to provide opportunities for a diverse group of licensees. We set up aggressive build-out requirements that will benefit consumers and small businesses everywhere. But I think we fell short on getting the rules right for small so-called designated entities, to give them a boost in the auction, and I hope we will reconsider some of the restrictions that we placed on them.

Unlicensed broadband services can also cover many underserved areas and hold promise for small providers. Unlicensed spectrum is free. It can be accessed immediately and equipment is relatively cheap. We are working to make more unlicensed spectrum available at higher-power levels.

There is also a lot more than Congress can do outside the purview of the FCC, such as providing adequate funding for RUS broadband loans and grants and properly targeting those loans and grants, providing tax incentives for companies that invest in broadband in underserved areas, promoting broadband in public housing, investing in basic science R&D, improving math and science education, and, of course, making sure that all of our children have affordable access to their own computer, because without a computer, broadband doesn't help.

We sorely need leadership like this Committee is showing today at all levels of government. It is time for a series of national broadband summits mediated by the Federal Government in partnership with the private sector to restore our place as the world leader in telecommunications. I look forward to working with you to maximize the availability of affordable, truly high-speed broadband services.

Finally, I would like to highlight an issue that Commissioner Copps mentioned. I know you both have expressed a lot of concern about the deplorable state of minority and female ownership of media assets. That is why I am encouraging the Commission to create an independent bipartisan panel to address these concerns. It is my hope that with your support and leadership, the Commission will do just that.

Thank you for inviting me to testify today.

[The prepared statement of Mr. Adelstein follows:]

STATEMENT OF JONATHAN STEVEN ADELSTEIN

STATEMENT OF
JONATHAN S. ADELSTEIN
COMMISSIONER, FEDERAL COMMUNICATIONS COMMISSION

BEFORE THE
U.S. SENATE COMMITTEE ON SMALL BUSINESS AND ENTREPRENEURSHIP

SEPTEMBER 26, 2007

Mr. Chairman, Senator Snowe, and members of the Committee, thank you for inviting me to testify about one of the most important infrastructure challenges confronting our Commission and the country: ensuring the ubiquitous deployment of affordable, high speed broadband to every small business that needs it. Deploying broadband infrastructure is critical to promote the economic, cultural, and social well-being of our country, particularly for small businesses and entrepreneurs, who drive so much innovation and economic growth.

Mr. Chairman, you have long recognized that broadband infrastructure is one of the best tools for promoting the entrepreneurial spirit that we have seen in our time. Its availability is fast becoming one of the key factors in the success or failure of our small businesses. I am deeply concerned about the speeds, prices, and availability of broadband services for American consumers and small businesses. To ensure that broadband is available and affordable, we must engage in a concerted and coordinated effort to restore our place as the world leader in telecommunications. This will require a comprehensive national broadband strategy that targets the needs of all Americans, including small businesses.

Mr. Chairman, I want to commend you for your leadership on this issue, not only through convening this hearing but also by serving as Co-Chairman of the Senate Democratic High Tech Task Force. Through these and many other efforts, you are drawing much-needed attention to the importance of promoting technological innovation and advanced telecommunications for providing good jobs and enhancing our standard of living. Senator Snowe, I also commend your outstanding leadership in promoting broadband for schools, libraries and health centers, as well as for all consumers, including those in rural areas.

The Role of Broadband in Promoting Economic Prosperity and Global Competitiveness

Small businesses play a driving role in creating jobs and developing new technologies. Over the past decade, small businesses have created two out of every three new jobs, employed forty percent of high tech workers, and produced far more patents than similarly focused large firms. Small businesses also purchase a massive amount of telecommunications services, spending approximately $25 billion each year, according to a recent Wall Street Journal report.

Unfortunately, the FCC collects little reliable data about extent of broadband services available to small businesses in the U.S., or the more general state of competition among providers of telecommunications services for businesses. In a report released at the end of last year, the U.S. Government Accountability Office (GAO) recommended that the Commission collect additional data to monitor competition and to assess customer choice through, for example, price indices and availability of competitive alternatives. GAO found that "without more complete and reliable measures of competition, [the] FCC is unable to determine whether its deregulatory policies are achieving their goals."[1]

The good news is that small businesses are voraciously integrating new services and features into their business plans. As I elaborate below, and as you will hear from the second panel, businesses of all sizes are increasingly tapping into broadband to reduce costs, increase productivity, and improve efficiency.

The bad news is that what little data that we have suggests that small businesses are starved for telecommunications choice. Many small businesses have only one choice of provider for broadband services, which deprives them of innovative alternatives and can result in higher prices. Even where there are competitive options, alternative providers rely heavily on inputs from incumbents, highlighting the importance of pro-competitive policies. GAO found that competitive providers serve, on average, less than six percent of the buildings with demand for dedicated access, leaving 94 percent of the market served by only incumbent providers.[2] These inputs are used not only by large businesses, but also by other communications providers, including independent wireless, satellite, and long distance providers that serve small businesses. It is noteworthy that the U.S. Small Business Administration Office of Advocacy recently commented that "[t]he combination of high prices and few alternatives creates an insurmountable burden to small carriers trying to conduct business in the telecommunications market."[3]

The lack of information about the small business market is particularly troubling because broadband creates economic opportunities that were previously unattainable, and the upside potential remains vast. Broadband can connect entrepreneurs to millions of new distant potential customers, facilitate telecommuting, and increase productivity. Much of the economic growth we have experienced in the last decade is attributable to productivity increases that have arisen from advances in technology, particularly in telecommunications. These new connections increase the efficiency of existing business and create new jobs by allowing new businesses to emerge, and spur new developments such as remote business locations and call centers.

Small businesses that have seized the initiative are witnessing tremendous growth. With broadband, you need not have a global marketing department to be accessible to the world. This capability is particularly potent for small businesses given that 52 percent are home-based. In this way, broadband is an extension of the entrepreneurial spirit that has characterized our country from its earliest foundations. Just as the Pilgrims used the Mayflower to reach the new opportunities in Plymouth Harbor and the 19th century pioneers relied on stage coaches and railroads to settle the western U.S., entrepreneurs are using broadband infrastructure to reach beyond their current horizons.

Since I have joined the Commission, I have traveled across the country and seen broadband technologies harnessed in ways folks inside the Beltway might never have imagined.

---

[1] GAO, *FCC Needs to Improve its Ability to Monitor and Determine the Extent of Competition in Dedicated Access Services*, p. 15 (Nov. 2006).

[2] *Id.* at 12.

[3] U.S. Small Business Administration Office of Advocacy, Comments in WC Docket No. 05-25, p. 8 (2007).

For example, at auction houses across the Midwest, entrepreneurs are using broadband technologies to conduct real time cattle auctions over the Internet. Ranchers from across the country can log in, watch real time video of the livestock and make purchases without leaving their ranches. These auction houses bridge remote locations, expand potential markets for livestock, and cut costs for ranchers to reach their customers.

Broadband can also unlock transformational opportunities through the cooperative nature of the Internet. Companies of all sizes are tapping into the power of the Internet to gather and develop ideas for new products, and to interact with and solicit the views of influential customers. The success of the on-line encyclopedia Wikipedia is just one example of how quickly open and accessible Internet-based models can outstrip their traditional predecessors. We are just scratching the surface of the opportunities that these technologies can bring.

As small businesses are increasingly empowered to use broadband in newer, more creative ways, the stage on which we all must compete is also evolving into a global one. New telecommunications networks are a key driver of this new global landscape. They let people do jobs from anywhere in the world -- whether an office in downtown Manhattan, a home on the Cheyenne River Indian Reservation, or a call center in Bangalore, India. This trend is a wake-up call for Americans to demand the highest quality communications systems across our nation, so that we can harness the full potential, productivity and efficiency of everyone in our own country. If we fail in this, be assured, our competitors around the world will take full advantage of it.

The Need for a National Broadband Strategy

We must do far more to give our citizens, our small business, and our communities the tools they need to succeed. We've made progress, many providers are deeply committed, and there are positive lessons to draw on. Yet, I am increasingly concerned that we have failed to keep pace with our global competitors over the past few years when it comes to the speeds, prices, and availability of broadband services.

For a long time, the U.S. was the undisputed world leader in communications technology. Yet, in recent years, we have tumbled from that historic position. Each year, we slip further down the regular rankings of broadband penetration. While some have questioned the international broadband penetration rankings, the fact is the U.S. has dropped year-after-year. This downward trend and the lack of broadband value illustrate the sobering point that when it comes to giving our citizens affordable access to state-of-the-art communications, the U.S. has fallen behind its global competitors.

There is no doubt about the evidence that citizens of other countries are getting a much greater broadband value in the form of more megabits for less money. A recent OECD report ranked U.S. 12th in broadband value. According to the ITU, the digital opportunity afforded to U.S. citizens is 21st in the world. For small businesses, those in rural areas, and low income consumers, the problems are often even more acute. This is more than a public relations problem. It is a major productivity problem, and our citizens deserve better. Indeed, if we do not do better for everyone in America, then we will all suffer economic injury. In this broadband world, more than ever, we are truly all in this together and we need to tap all of our resources.

Some have argued that the reason we have fallen so far in the international broadband rankings is that we are a more rural country than many of those ahead of us. While this is debatable, even if it is the case, we should redouble our efforts and get down to the business of addressing and overcoming this challenge.

I am concerned that the lack of a comprehensive broadband communications deployment plan is one of the reasons that the U.S. is increasingly falling further behind our global competitors. Virtually every other developed country has implemented a national broadband strategy. This must become a greater national priority for America than it is now. We need a strategy to prevent outsourcing of jobs overseas by promoting the ability of U.S. companies to "in-source" within our own borders. Rural America and underserved urban areas have surplus labor forces waiting to be tapped. No one will work harder, or work more efficiently, than Americans. But too many are currently without opportunities simply because their current communications opportunities are inadequate to connect them with a good job. That situation must improve.

The Elements of a National Broadband Strategy

A true broadband strategy should incorporate benchmarks, deployment timetables, and measurable thresholds to gauge our progress. We need to set ambitious goals and shoot for affordable, truly high-bandwidth broadband. We should start by updating our current anemic definition of high-speed of just 200 kbps in one direction to something more akin to what consumers receive in countries with which we compete, speeds that are magnitudes higher than our current definitions.

We must take a hard look at our successes and failures. We need much more reliable, specific data than the FCC currently compiles so that we can better ascertain our current problems and develop responsive solutions. The FCC should be able to give Congress and consumers a clear sense of the price per megabit, just as we all look to the price per gallon of gasoline as a key indicator of consumer welfare. Giving consumers reliable information by requiring public reporting of actual broadband speeds by providers would spur better service and enable the free market to function more effectively. Another important tool is better mapping of broadband availability, which would enable the public and private sectors to work together to target underserved areas. Legislation under consideration by leaders in both the Senate and the House would enable us and other agencies like the Census Bureau to make enormous progress on this front.

We must redouble our efforts to encourage broadband development by increasing incentives for investment, because we will rely on the private sector as the primary driver of growth. These efforts must take place across technologies, so that we not only build on the traditional telephone and cable platforms, but also create opportunities for deployment of fiber-to-the-home, fixed and mobile wireless, broadband over power line, and satellite technologies.

We must work to promote meaningful competition, as it is the most effective driver of innovation, as well as lower prices. Only rational competition policies can ensure that the U.S. broadband market does not devolve into a stagnant duopoly, which is a serious concern given that cable and DSL providers now control approximately 96 percent of the residential broadband market.

The Commission must also ensure the vitality of universal service as technology evolves. With voice, video, and data increasingly flowing to homes and businesses over broadband platforms, we've got to have ubiquitous high speed networks to carry these services everywhere, so that small business owners in all parts of our country can participate in this global economy. Universal service must evolve, as Congress intended, to cover broadband services sooner rather than later. As elaborated upon below, we must also promote spectrum-based services that can play such an important role spurring both competition and greater availability of these services.

There also is more Congress can do, outside of the purview of the FCC, such as providing adequate funding for Rural Utilities Service broadband loans and grants, and ensuring RUS properly targets those funds; establishing new grant programs supporting public-private partnerships that can identify strategies to spur deployment; providing tax incentives for companies that invest in broadband to underserved areas; devising better depreciation rules for capital investments in targeted telecommunications services; promoting the deployment of high speed Internet access to public housing units and redevelopments projects; investing in basic science research and development to spur further innovation in telecommunications technology; and improving math and science education so that we have the human resources to fuel continued growth, innovation and usage of advanced telecommunications services; and, of course, we need to make sure all of our children have affordable access to their own computers to take full advantage of the many educational opportunities offered by broadband.

What is sorely needed is real leadership at all levels of government, working in partnership with the private sector, to restore our leadership in telecommunications. This Committee's attention to this issue is exactly the kind of effort that is needed. I also continue to believe that we need a National Summit on Broadband -- or a series of such summits -- mediated by the federal government, including Congress, the Executive Branch and independent agencies, and involving the private sector, which could focus the kind of attention that is needed to restore our place as the world leader in telecommunications.

Wireless: A Critical Source of Broadband Services

One of the best opportunities for promoting broadband, and providing competition across the country, is in maximizing the potential of spectrum-based services. The Commission must do more to stay on top of the latest developments in spectrum technology and policy, working with both licensed and unlicensed spectrum. Spectrum is the lifeblood for much of this new communications landscape. The past several years have seen an explosion of new opportunities for consumers, like Wi-Fi, satellite-based technologies, and more advanced mobile services. We now have to be more creative with what I have described as "spectrum facilitation." That means looking at all types of approaches – technical, economic or regulatory – to get spectrum into the hands of operators ready to serve consumers at the most local levels possible.

Of course, licensed spectrum has and will continue to be the backbone for much of our wireless communications network. We are already seeing broadband provided over satellite, new wireless broadband systems in the 2.5 GHz band, and the increasing deployment of higher speed mobile wireless connections from existing cellular and PCS providers.

During our review of the bandplan in advance of the auction last year of 90 MHz of new spectrum for the Advanced Wireless Service, I pressed for the inclusion of smaller blocks of licenses. I thought that smaller license blocks would improve access to spectrum by those providers who want to offer service to smaller areas, while also providing a better opportunity for larger carriers to more strategically expand their spectrum footprints. According to the U.S. Small Business Administration Office of Advocacy, which conducted a roundtable to discuss FCC spectrum policy, "small businesses identified the size of the license areas as the single greatest regulatory barrier to providing wireless service."[4] Not surprisingly, our decision to adopt smaller license blocks was well received by a number of carriers and manufacturers.

The Commission to some extent used the historic opportunity in the upcoming 700 MHz auction to facilitate the emergence of a "third" broadband platform. This is the biggest and most important auction we will see for many years to come. While the Commission recently adopted auction rules that reflect a compromise among many different competing interests, I am hopeful that there will be opportunities for a diverse group of licensees in the 700 MHz auction and that our more aggressive build-out requirements will benefit consumers across the country. We also put in place a new approach to spectrum management by adopting a meaningful, though not perfect, open access environment on a significant portion of the 700 MHz spectrum. This decision represents an honest, good faith effort to establish an open access regime for devices and applications that will hopefully serve consumers well and create opportunities for small providers for many years to come.

I have been disappointed, however, with the way that the Commission has handled its designated entity (DE) program. The bidding credits made available through this program can be a potent means of getting spectrum into the hands of small businesses and entrepreneurs. Yet, the Commission has missed the chance, time and again, to craft rational DE rules. So, I was again disappointed that, in the 700 MHz proceeding, we lost an opportunity to provide crucial bidding credits to designated entities that wholesale fully built-out network services. I think it is essential that we revisit our policies in this respect to ensure that all bidders have opportunities to bid, particularly where wholesale service is a compelling option for new and diverse providers.

Beyond the 700 MHz auction, there are other important opportunities for small businesses as both consumers and providers of broadband services. Unlicensed broadband services are an intriguing avenue for many underserved communities because unlicensed spectrum is free and, in most rural areas, lightly used. It can be accessed immediately, and the equipment is relatively cheap because it is so widely available. I have also worked closely with the Wireless Internet Service Provider (WISP) community, which has been particularly focused on providing wireless broadband connectivity in rural and underserved areas.

---

[4] Letter from Eric E. Menge, U.S. Small Business Administration Office of Advocacy, to Ms. Marlene Dortch, FCC, WC Docket No. 06-150, p. 1 (Dec. 7, 2006).

But we can always do more for rural WISPs and other unlicensed users. I have heard from operators who want access to additional spectrum and at higher power levels. And the Commission has been doing just that. We have opened up 255 megahertz of spectrum in the 5 GHz band – more spectrum for the latest Wi-Fi technologies – and are looking at ways to increase unlicensed power levels in rural areas.

I also have pushed for flexible licensing approaches that make it easier for community-based providers to get access to wireless broadband opportunities. We adopted rules to make spectrum in the 3650 MHz band available for wireless broadband services. To promote interest in the band, we adopted an innovative, hybrid approach for spectrum access. It makes the spectrum available on a licensed, but non-exclusive, basis. I have spoken with representatives of the Community Wireless Network movement, and they are thrilled with this decision and the positive impact it will have on their efforts to deploy broadband networks in underserved communities around the country.

We have also made spectrum available in the 70/80/90 GHz band for enterprise use. While you may not be familiar with this spectrum block, it can be used to connect buildings with gigabit-speed wireless point-to-point links for a mile or more. Instead of digging up streets to bring fiber to buildings, licensees can set up a wireless link for a fraction of the cost -- and the spectrum is available to anyone holding a license. While others supported an auction, I successfully argued against them in this unique case, because I was concerned that auctions would raise the price of access and shut out smaller licensees. In fact, one company now is installing five links in my home state of South Dakota. The links will be used for a number of city services, including public works, police and fire departments, as an alternative to fiber.

Conclusion

In order to maintain a vibrant environment for our nation's entrepreneurs and small businesses, we need to maximize the availability of affordable, truly high-speed broadband services. I look forward to hearing from the next panel of witnesses and working with you all to create a comprehensive policy framework that advances that goal. Thank you for your leadership on this issue, and for inviting me to testify today.

Chairman KERRY. Well, thank you both for important testimony.

We now have four statements today, mine, the Ranking Members, and both of yours, that describe the problem, and both of you have described it succinctly, eloquently, forcefully, and compellingly. So the question is, I mean, who is supposed to do this? Why is this not happening? What is the problem here?

Commissioner COPPS. Well, I think first of all is the lack of a strategy. Number two is the lack of good information so people can understand the problem. But I think as important as anything has been the mindset that we have been working under for the last several years—not to worry about it. The marketplace will take care of this. The market is going to provide ubiquitous broadband. It is going to protect the public interest in media, too. Nothing else is needed.

While we all revere the marketplace, which is the locomotive of our system and should always be in the lead, there are some things that are not getting done, cannot get done by themselves. You can go back, as I said, to our early history, building the infrastructure that we needed to places where it had to go but the private sector didn't see an immediate profit by going there.

So we need to cooperate. We need to innovate. We need to learn what municipalities are——

Chairman KERRY. What do you think the most significant step would be, legislative structure, executive order, or an economic incentive? What is going to have the biggest return here in terms of people saying, wow, now we can go do this?

Commissioner COPPS. I think a committment from on high saying that this is the infrastructure challenge of the first part of the 21st century. We have always built America and kept it great by keeping up with infrastructure. We have to do that with our physical infrastructure, and broadband is the highway and the byway and the ports and the canals and the railroads of the 21st century. Without it we are going to be left behind. Then people will pay attention and then we can come in and do all this——

Chairman KERRY. Didn't we set that goal? Didn't the President set that goal in 2004?

Commissioner COPPS. Well, a goal is always welcome, but a goal has to be accompanied by a strategy and a strategy has to be informed by tactics, and that is where we have fallen down.

Chairman KERRY. Again, let me re-ask it. What tactic do you think would have the greatest impact? I mean, do we need to create some huge tax credit or incentive for rural investment? Do we need to create grants for rural investment?

Mr. ADELSTEIN. I think we need a comprehensive plan. I laid out today a comprehensive plan which involves both legislation and leadership on the national level, as well as action by the FCC. The Telecommunications Act did envision this. It talks about advanced services five times in Section 254.

Chairman KERRY. The Telecommunications Act envisioned that we were going to have local Bell Telephone Companies competing in the marketplace and frankly, the regulators didn't regulate. Mr. ADELSTEIN. That is right. We basically gave up on it.

Chairman KERRY. Absent some enforcement, nothing happened Mr. ADELSTEIN. We gave up on competition. Competition drives deployment like nothing else. The vision of the Act was competition. Now, it is not Congress's fault that the FCC gave up on the job and the marketplace didn't work very well. Now we have consolidation and lack of competition, and as Free Press's testimony indicates, competition should be the biggest driver of prices. Prices are shooting up. There are no alternatives for these small businesses. We need a coordinated plan from the highest levels.

I mean, one way to start is a national summit on broadband. Why don't we have this kind of leadership where we all gather, private sector, public sector, Congress, the Executive branch agencies including us, NTIA, all the way down the line. That brings everybody together. I also laid out a comprehensive plan here today— tax credits. You need grants. You need universal service. But you also need the FCC to promote competition policies and create the incentives to invest.

Chairman KERRY. You talked about more megabits for less money in Europe. What were you referring to?

Mr. ADELSTEIN. In Europe and Japan, all around the world, the OECD data shows that we are paying more for less. In Japan, you get——

Chairman KERRY. Why are we paying more for less, is that because of lack of competition?

Mr. ADELSTEIN. Well, it is lack of competition. In some cases in these countries, actually, have regulated monopolies which are resulted in faster speeds at lower prices. We pay seven times as much as Japanese consumers for lower speeds, and they have a more regulated environment. So we have this duopoly here, but apparently a duopoly isn't sufficient. A lot of small businesses don't have access to a cable provider at all, so they only have one choice because cable doesn't go to the business areas. We see that they are trying to compete, but there has been an attempt to squeeze and destroy the CLECs and they are in need of protection to have regulatory stability.

Chairman KERRY. Who do you believe could be the critical players at that summit?

Mr. ADELSTEIN. Well, I think it has to come from the top on down. I think that leaders from the Executive branch to the Congress, the leaders of the committees, yourself included, of course, and this Committee as well as the Commerce Committee. I think that the private sector, all of the major leaders from the very small providers and the CLECs to the very largest national providers need to all come together to talk about making this a national priority and set goals and benchmarks. It is one thing to say you are going to get there by a certain date, but what are the exact benchmarks by which you get there? How do you measure that? What is the data that you need to get there? We need to all come together with that kind of leadership. Knocking heads together could make a difference.

Commissioner COPPS. But meanwhile, there are concrete things we can do. We talked about better data gathering and analysis, but the joint board is talking right now about including broadband specifically in universal service. I think we have the authority to do that under the Communications Act. We used universal service to get plain old telephone service out to all of our citizens, or most of them. That was the pots. Now we have got the pans, the pretty awesome new stuff, and we ought to find a way to get the pans out, as well as the pots to all Americans and we are not doing it, and this is a fix that could be made in the near-term future. So we would be at least taking one fairly significant step.

Chairman KERRY. You talked about the past, we have great examples of this: for example, electricity in America and the TVA and the effort to say we are going to get electricity out to every home in America. Is there a sense that the Internet ought to be, at least until broadband is universal, treated as more of a public commodity?

Mr. ADELSTEIN. I think so. We should make broadband the dial tone of the 21st century. The Farm Bill in 2001 did take RUS from being just a telephone system to a broadband system. I talked to somebody last night from RUS, they are having more applications coming in than they can fund this year, great applications coming in. So that is one step. But it has to be like the National Highway System, as well. If it weren't for Eisenhower making the commitment, we wouldn't have the highway system we have today. That vision back in the 1950s needs to be happening now, I think, for the Internet system.

Commissioner COPPS. You ask about how we are treating the internet. We are not even treating it as a telecommunications service here in the 21st century. We have spent all this inordinate amount of time at the FCC deciding that, oh, this isn't telecommunications, this is an information service so none of the consumer protections, universal service, privacy obligations apply to it. Here we go in with all of this wonderful new technology, all of the awesome opportunities it has for the future of this country in the 21st century and we don't even apply the simple protections that applied to plain old telephone service in the last century. That is a shame and a sham.

Chairman KERRY. Senator Snowe.

Senator SNOWE. Thank you, Mr. Chairman.

I thank both of you for your very powerful statements on these issues that are clearly are resonating and reverberating across this country. There are a multiplicity of problems, without a doubt, the President did set a goal in 2007, and looking back, I remember thinking we would have plenty of choices when it came to purchasing the broadband carrier. The more choices around, the more the price will go down. The more the price goes down, the more users there will be. And with more users, it becomes more likely that America will stay on the competitive edge of world trade.

Obviously, that hasn't occurred and it just can't happen magically. We have to develop complementary remedies between the Congress, the FCC, and the Administration. I think the idea of a broadband summit is an excellent way to start crafting a national strategy where each branch of government understands exactly what it is required to do. I was asking my staff last night who does what? It is critically important that each branch of the government understand their role, and I'm concerned that they don't. This is a multifaceted issue, and obviously you have to orchestrate a comprehensive strategy, and if it is important to America's economy then clearly there should be a national broadband policy. Everybody has discussed, but it clearly hasn't happened.

There are several issues that I would like to explore. One is on the use of the Universal Service Fund for broadband services, the high-cost fund. Is it clear, Commissioner Copps, as to whether or not you can use the Universal Service Fund for the support of broadband deployment, because you have had a reclassification of broadband services as information rather than telecommunications service. Is that a legal hindrance to using the fund?

Commissioner COPPS. No, I don't think it is a legal hindrance. Certainly it would be doable under the ancillary authority of title 1, if nothing else. But I think clearly we have not only the authority to do it, but the charge from Congress to get advanced telecommunications to all of our citizens.

Senator SNOWE. Last year, Senator Stevens and other Members of the Commerce Committee worked on the universal service issue. Five hundred million dollars was included in the Universal Service Fund to help deployment in rural areas. Do you think that this funding has had an impact?

Commissioner COPPS. I think that is helpful. I think in the long run, to get broadband deployed around the country is going to be a very expensive exercise. We are looking right now at trying to get a public safety broadband system established through the 700 megahertz auction and that is going to be billions of dollars just to do that.

Senator SNOWE. Commissioner Adelstein, you made a good point about tax incentives and grants, could they be supported by supplementing the Universal Service Fund, and would it help with respect to this type of deployment?

Mr. ADELSTEIN. I think that is right. I mean, a lot of people say the reason we are falling behind is we are rural. I am not sure that is entirely supported by the evidence, but to the extent that is true, and you know the rural parts of Maine, you look at Western Massachusetts, we do have a problem in rural areas. So if that is the problem, why don't we redouble our efforts? Why don't we focus broadband on that and the access advanced services in section 254 where we have that authority? We need to do tax credits to encourage areas where the market isn't serving, and the RUS program is, I think, really doing a great job of getting broadband out and it needs to do even more. It needs to be fully funded, as well.

Senator SNOWE. What about the special access issue? Is that a major factor that will help to promote competition? There are a lot of small companies that are dependent on the Bells for the infrastructure and access. In many cases it is only one company that small companies rely on and their prices are high and becoming even more costly. I know that there is a decision pending before the FCC, but would that help?

Commissioner COPPS. It is pending. I think if we can get it right, it would help. We had a GAO study recently that pointed up the problems that attend special access. There is a lot of money involved in it, $15 or $16 billion charged by the big phone companies, and about 94 percent of the country's enterprise buildings are reached only by the big ILECs. Is that a drain and a hindrance to small business? The GAO thought so, and I think so, too. So we are under an obligation to get this done, kind of a self-imposed one by the first of October. That is 4 days away. I haven't seen the item yet.

Mr. ADELSTEIN. Certainly, that is right. Businesses, long-distance providers, and wireless all rely on special access. Customers say they don't have any competition, that the earnings by the Bells are excessive. The GAO report that Commissioner Copps referred to found that there are competitors in only 6 percent of the market. Ninety-four percent of buildings are only being served by local incumbents. These are buildings where small businesses are located and large businesses as well. So this impacts everybody. It ripples throughout the system. Businesses, hospitals, governments all pay more than the market might otherwise determine, if it were truly competitive.

And if you think about a new competitor coming in, like a new national wireless system we are hoping under the 700 megahertz auction, every little node they set up, every tower they set up is going to have to use special access to connect to the network, and so we have to make sure that we get this right.

Senator SNOWE. What about broadband mapping? Would that be helpful to pass mapping legislation? Would that help us know exactly where broadband has been deployed and where it hasn't, and is this something the FCC is already undertaking?

Commissioner COPPS. No. But it would be immensely helpful. It is something that FCC should long ago have done and long ago provided to you and provided to companies around the country. Now, thank goodness we have all of these exercises, Connect Kentucky and Connect America generally, and a number of States are doing this and I applaud that. But this is something if we had a national strategy the FCC would have been charged to complete a long time ago. We shouldn't be messing around with this in 2007, finding out who has got what.

Mr. ADELSTEIN. And we should be mapping—I think Connect Maine will help, as Connect Kentucky did. I have this map of Massachusetts. You look at the FCC's data and compare it to that, the FCC says you have broadband everywhere in Massachusetts, but you look at all those red areas in Western Massachusetts and that is not the case at all. So the FCC's data is clearly inadequate. Our maps are a disgrace. They are not adequate to give us a real picture of what is happening.

Chairman KERRY. Is that the John Adams Institute or——

Mr. ADELSTEIN. Yes. That John Adams map there shows all those red areas with no broadband, but the FCC's map, their different color codes show that you have broadband everywhere in Massachusetts. So our mapping is completely inadequate.

Now, it is not that hard to do. I was in Chicago last week and there was a small businessman, Willie Cade, who owns PC Rebuilders and Recyclers. He, on his own, came up with a program, his little small business, that mapped all of Chicago, everything that the major providers are providing in Chicago, and you can see, as a matter of fact, there tends to be more service in the higher- income areas than in the lower-income areas, all mapped out. He managed to mine the data from publicly available information that the providers have on their own Web sites. So why can't we do it?

If a small business in Chicago can do it, why can't the Federal Government do it?

Senator SNOWE. Well, that is a very good question. Why can't we?

Mr. ADELSTEIN. I think we can. I think we should. Legislation would be helpful, but the FCC must undertake, I think, a better role. I talked to Chairman Martin this morning and I think he shares the commitment to improving the data that we get. We have a proceeding that is pending right now. We need to make sure that we have good mapping as a part of that and make sure that we ascertain small business and what kind of availability small businesses have.

Senator SNOWE. So that is something that you think that the FCC will pursue?

Mr. ADELSTEIN. I do think so. We have a pending open proceeding right now. Just this morning we discussed the need to ensuring that we get better data. We are going to work very hard to make sure that it is as strong as it can be. We would like to work with you, as well, to get your input.

Senator SNOWE. Thank you.

Chairman KERRY. Remember the old statement, trust but verify.

Senator SNOWE. Yes. Exactly. But I appreciate it, because it is clear to me that we have a lot to do with those branches and with the agency. We have to figure out how to corral all of this and just have a clear strategy for the future and pursue it aggressively. Thank you very much. Thank you, Mr. Chairman.

Chairman KERRY. Thank you very much, Senator Snowe.

I would like to reference this map for the moment a moment. It is up in the back here, and Senator Corker, I will recognize you in just 1 minute. I want to point out that the red areas are the entire areas and towns that have no access at all, and yet Massachusetts is ranked number four in the country. This is why this is important. The orange areas represent where broadband is available in a very limited amounts. The yellow shows areas that have only one broadband provider. As you can see, it is a complete monopoly—no competition—therefore pricing is not competitive.

A duopoly is where you have two broadband providers and is shown in blue. Two is not sufficient in many people's judgments. And you have only this tiny area around Boston, the sort of greater Boston area there, where you actually have three or more broadband providers and real competition. So most of the State of Massachusetts doesn't have real competition (more than three providers) which is an extraordinary statement about where we stand with broadband penetration.

[The broadband availability map referenced above follows:]

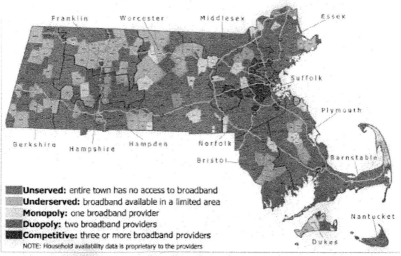

Broadband Availability in Massachusetts Municipalities
June 2007

Unserved: entire town has no access to broadband
Underserved: broadband available in a limited area
Monopoly: one broadband provider
Duopoly: two broadband providers
Competitive: three or more broadband providers
NOTE: Household availability data is proprietary to the providers

John Adams Innovation Institute

Senator KERRY. Senator Corker.

Senator CORKER. Mr. Chairman, thank you for pulling together this hearing. I was in 58 counties of the 95 that we have in our State during August recess and broadband is a big issue, especially in the rural areas. In our own State, a lot of the municipalities, I know when I was mayor we put in a 96–fiber line around our city to create some competition and I know other cities are doing the same in our State. Some of the rural areas obviously are applying for grants to do the same kind of thing. But it is an issue, no doubt.

I do wonder, I hear us talking about Federal mapping and all of that. I know that States are also engaged in many cases. I know we have a gentleman, Mr. Mefford, who is actually involved in the State of Tennessee right now connecting our State and is going to be part of the second panel which I am going to miss, but what role do you as Commissioners see at the State and local level?

It seems like that we have a tendency here to want to Federalize everything and I know there are a number of activities that are taking place in States across the country and I would love for you all to make comment on that.

Commissioner COPPS. Well, I think it is an important question and I think probably we have actually Federalized too much in the way we have approached telecommunications policies and taken away authority from the States on a lot of the consumer and other issues. The franchising exercise that we went through was another example of that. So we have to get back to the kind of a balance that I think the Telecommunications Act of 1996 envisioned between Federal and State authorities.

There are some things, I think, that are obviously more efficiently done in one venue than another, and I think getting baseline data on broadband and deployment and knowing who has it and measuring the speeds and all that is a perfectly legitimate exercise for the Federal Communications Commission and is something we should have done long ago. You know, a lot of States don't have the resources to do that and a lot of the States don't have the "connect" initiatives that many States are developing right now.

This is a national problem. It is a national challenge. It is a global competitive challenge to our small businesses and we have to treat it that way and use all of the resources we have, Federal, State, local. We need to innovation and we need to learn from what various States and localities are doing.

Mr. ADELSTEIN. I think it really is a partnership that we need to do with our State and local colleagues. I was meeting with the mayor of Fort Wayne recently who has done an incredible job of getting Fort Wayne wired, working with providers, but what does that mean for Gary? What does that mean for South Bend? It's great for Fort Wayne, and they are going to get business that other cities won't get, but what about having a national system and working with innovative mayors like that, working with the States that are doing things like Connect Maine or Connect Kentucky? Where is Connect South Dakota? Are they going to get left behind if they don't get it together? Can't we all have similar maps so that we have a uniform national vision of this?

I think we can learn a lot from what the State and local governments are doing. As a matter of fact, when it comes to the national summit, I wanted to say one of the ideas is you really want to include State and local governments in that. If need be, Congress itself could convene such a national summit—it doesn't have to come from the Executive branch—and invite Executive branch partners to come in along with State and local governments to talk about what are some of the great things that are going on in places like Fort Wayne, and why can't we do that nationally. Those cities that have good visionary leadership shouldn't have an unfair advantage over those that, unfortunately, for whatever reason, don't have leaders that are so focused on telecommunications.

Senator CORKER. You know, we have had people in our office. I find the testimony today somewhat interesting. I think, in particular, Mr. Copps is kind of a cranky testimony, if you will, and you are involved, it seems, the FCC is sort of the centerpiece at the Federal level being involved in these kind of things.

We have had people in our office talking about the auctioning of some of these spectrums that you all are talking about that say that they are perfectly willing to connect every—make sure that every home in America has access to broadband if they can just get these spectrums bid appropriately so that they have the opportunity to do that. I would love for you all to comment, because it sounds like there are some things that you readily have available to solve some of these problems.

Commissioner COPPS. Well, I am old, and I am cranky, and I have been in this town for 37 years now and——

Senator CORKER. You wear it well.

Mr. COPPS [continuing]. Dealing with this problem of small and medium-sized enterprise for much of that time, so that is why I get a little bit impatient.

Yes, the rules and the procedures we establish for our auctions are very important. You have to look at each case that comes along. I mean, some people want to get that spectrum in very unconventional ways that sometimes may be in contravention of the statute or maybe go around auctions or something like that, so you have to look at each of those cases, but we have to be innovative. That is why Jonathan and I were concerned on the 700 megahertz auction that we weren't more innovative to encourage more participation and have open access, a wholesale model, to allow some competition in at least this one part of this one piece of spectrum. Let us try something different and see if it works. Yes, we have the authority to do that, and we should be doing a lot more of it than we are.

Senator CORKER. I mean, here we are testifying before a Senate Committee. Why don't you tell us why you are not doing that? It seems like to me that you have the tools at your disposal at the FCC truly on these spectrum auctions to solve this problem——

Commissioner COPPS. I think we do. But number one, I just observe that we are two out of five people, so we don't necessarily command a majority for everything that we want to do.

Chairman KERRY. They don't have the votes. The Commission is appointed——

Senator CORKER. I understand there are five, but I can't imagine—I would love to get some of the other Commissioners up here then, Mr. Chairman, and talk about it. But I would sure love for you, since you seem a little perturbed about it, for you to air why that is not occurring.

Commissioner COPPS. It is not occurring for all of the reasons that I have tried to explain this morning, beginning with the lack of a national strategy. We don't have that charge from on high to get this job done. We don't have the charge saying, this is the most important infrastructure problem our country faces. That charge would say: go and use the authority you have and get it done, and if you don't have the authority, come back here and get some more.

It is either going to be a priority or it is not going to be a priority and we are not treating it as a priority, and to me, it is the central infrastructure challenge that we face right now. If we don't do this, small business is going to suffer. Minorities are going to suffer. Rural America is going to suffer. And the country as a whole is going to suffer. It is a job that is not getting done and——

Senator CORKER. Again, I don't want to create acrimony here, but I just have people come in our office representing companies from around the country that feel like they could solve this problem. You all are two of the five Commissioners apparently that could affect that and I think that is an area for us to begin——

Commissioner COPPS. I think that is true, but in the deregulated environment in which we live, which is the environment that a lot of these companies pushed for, we were told that if we would deregulate the job would get done. We deregulated. The job didn't get done.

Mr. ADELSTEIN. I think I know one of the companies you are referring to. You are talking about getting wireless spectrum into use, and that is something that I talked about. I think you are exactly right on. There are opportunities out there to do it. Now, why didn't that happen? It is a very good question. The company perhaps is M2Z that you are talking about. This is a company that had a proposal for nationwide use of a certain area of spectrum that is now underutilized. They argued under section 7 of the Act that says we are supposed to get new services and new technologies approved or decided up or down within a year.

Now, they put forward a proposal, and it was a year before we even acted on it. We didn't even have the opportunity to vote on it or anything because nothing came before us for a whole year. Finally, we just put an NPRM out like the day before the year expired so we wouldn't be exposed in court, but why didn't we do it quicker? What are we waiting for before we even put out a notice asking what we should do about something?

Here this private company did identify, I think helpfully, that there was some underutilized spectrum and they wanted to do something with it. Well, whether you like what they want to do or not, why don't we find a way to get that spectrum into use, get them or somebody else using it, auction if off, get it moving, get that out for notice and get the auction up and running. I couldn't agree with you more. We need to be doing that. We need to look at every inch of spectrum we have and try to pack more data on it. Here is an example of where we didn't do our job well. We didn't really comply with the spirit of trying to get things done in a year and it is frustrating a little bit. It is making me age prematurely

Senator CORKER. I appreciate the time, and Mr. Chairman, I would just say——

Chairman KERRY. No, that is a very legitimate and very important series of questions. I think it does—clearly it begs the question that is on the table.

Senator CORKER. And I think that before we get involved in mapping and a Federal initiative and all that, I think there is an entity here that with some degree of innovation within its own ranks could go a long way toward solving this problem without—in a way that, candidly, is not something that would use a lot of Federal resources. I mean, you have spectrum. We have a need. You all have the ability to auction that spectrum in a way that creates universal access if you so decide, and I would just urge the Commissioners to maybe come back and talk with us about ways of making that happen.

But again, thank you for this, and I had no idea I was going to ask even these types of questions. It really came because Mr. Mefford is wiring our State and I wanted to pay tribute to him, but thank you for this testimony.

Chairman KERRY. No, we appreciate it. It is good to get everybody's crankiness out on the table.

[Laughter.]

Chairman KERRY. Thank you, gentlemen. We appreciate it.

We are going to go right to the second panel. We are under a little bit of time pressure here, so if we could just have a seamless transition, that would be terrific.

And I think, Senator Corker, it would be really worthwhile to get the other Commissioners in and have this conversation with them. I will do that. We will do that.

Senator CORKER. Thank you.

Chairman KERRY. Thank you very much. We appreciate it.

So Ben Scott from the Free Press, policy director, Brian Mefford, Doug Levin, and Scott Wallsten. If you could each summarize your testimonies in 5 minutes or less, that will help.

Mr. Scott, do you want to start, and we will just run down the line. Just identify yourself for the record and proceed.

STATEMENT OF BEN SCOTT, POLICY DIRECTOR, FREE PRESS, WASHINGTON, DC

Mr. SCOTT. Thank you, Mr. Chairman, Senator Snowe. Thank you for the opportunity to testify today. I am the policy director at Free Press. We are a public interest organization with over 350,000 members. We are dedicated to public education and consumer advocacy on communications policy.

Many of my members are small businesses and their interest in broadband could hardly be a higher priority. For them, broadband is a make or break technology. Many are E-commerce outfits, but almost all of them use the Internet to place orders, track inventory, or market products.

Unfortunately, a lack of competition in the broadband market has led to high prices and slow speeds for these small business connections. This has been going on for quite some time, threatening to stunt innovation and endangering our global competitiveness, as both of the Commissioners pointed out. I share their view that this is a very serious problem.

Increasingly, our small businesses are competing with similar enterprises overseas and we stand at significant disadvantage. A recent Small Business Administration study of broadband prices showed that small businesses in States like Massachusetts and Maine are likely to pay $40 or more for a six-megabit connection to a consumer-grade cable modem. Their competitors in Japan are paying the same price for 100 megabits. This 15-fold speed advantage translates into more goods, better services, and higher efficiency, and it is not just the Japanese that have the edge.

According to a study by the OECD of higher-quality enterprise-class broadband services, the United States once again pays far more than other nations for far less. What is available in Denmark for $350 to small businesses costs

$2,500 here at home. Now, I believe as much as the next guy in the power of the American entrepreneurial spirit, but the head start we are giving our global competitors is taking it just a bit too far.

So what do Asia and Europe have that we don't? They have competitive markets. They have competition that drives prices down and speeds up and we don't, and it is not hard to see the results.

In our study of this problem, we noticed how few small businesses actually subscribe to the high-end broadband services that best suits their interests. Most get by with a lower standard $40 consumer-grade broadband product. Only a fraction subscribe to enterprise-class services that could supercharge their businesses. According to the SBA survey from 2004, only 4 percent of small businesses were buying these high-end connections—4 percent. Even if we generously assume that since 2004 that number has tripled, that is just over 10 percent of our small businesses that are getting what they need.

The simple reason is high prices. That same SBA survey showed that these high-quality connections cost over $700 a month. The kind of competition necessary to bring those costs down is nowhere on the horizon. Meanwhile, the big phone companies are over at the FCC using their political muscle to push out these competitors.

Right now, the FCC is considering a number of critically important regulatory choices, including changes in so-called special access and network sharing policies that govern business class broadband. Wrong decisions could result in even higher prices for small business.

Another free market policy that is critical to small business is network neutrality. Small businesses depend on the Internet for E- commerce and they need net neutrality to protect the free market, ensuring that no large companies have unfair advantages. One of my members is a small business owner from Washington State who wrote me and captured this issue in a nutshell. He wrote, "I am the founder and CEO of a Web-based startup, so my life is dramatically affected by net neutrality. We will be competing against many major companies, so the possibility of a large ISP having the option of routing my traffic to a second-tier network is chilling, to say the least."I want to thank both of you, Senator Kerry and Senator Snowe, for your leadership on this critical issue, because to meet the needs of this CEO and others like him, my recommendation is that this Committee undertake a sweeping inquiry into broadband policies that affect small businesses in particular.

To begin, we need to improve our knowledge of the small business market. Currently, no Federal agency is consistently studying this problem. It seems to me we can't fix problems we don't measure, and since the SBA has already begun to conduct surveys of small business broadband, I think they ought to proceed, in co-operation with the FCC.

But above all, we need competition policy to drive down prices, accelerate speeds, and deliver better value to American small businesses. That means fostering more competition with innovative new technologies, like in the spectrum auction, but it also means forcing entrenched monopolies to open their networks to competitors. That is the key point that is holding up action at the Commission.

In the short term, I recommend moving forward on a variety of progressive policies which I outlined in detail in my written statement. These include opening the TV white spaces for unlicensed wireless use; protecting the rights of local government to offer broadband services; transitioning Universal Service Programs from dial tone to broadband; safeguarding the Internet's free market for goods, services, and speech through net neutrality rules; and finally, opening incumbent networks to unleash competitive forces.

In my view, this is a paradigm shifting moment for American telecommunications. It is an imperative that we choose wisely.

Thank you for your time and attention, and I do look forward to your questions.

[The prepared statement of Mr. Scott follows:]

Testimony of

Ben Scott
Policy Director
Free Press

on behalf of

Free Press
Consumers Union
Consumer Federation of America

before the

United States Senate
Committee on Small Business & Entrepreneurship

Regarding

Improving Internet Access to Help Small Business
Compete in the Global Economy
September 26, 2007

Free Press
Massachusetts Office
40 Main St., Suite 301
Florence, MA 01062
(413) 585-1533

Free Press
Washington Office
501 3rd St, NW, Suite 875
Washington, DC 20001
(202) 265-1490

SUMMARY -- TESTIMONY OF BEN SCOTT, FREE PRESS -- SEPTEMBER 26, 2007

Free Press[1], Consumers Union[2], and Consumer Federation of America[3] appreciate the opportunity to testify on the state of the broadband market for small businesses. Few issues have a more direct path to economic growth at the center of our entrepreneurial economy. Unfortunately, a lack of competition has led to high prices and slow speeds for small business broadband connections, threatening to stunt innovation and endangering our global competitiveness. Our primary policy goal must be to increase competition in the broadband Internet service provider (ISP) marketplace.

Policies that create a healthy broadband market are critical for our small business economy. To begin, competitive ISPs are often small businesses. The competitive ISP industry has dramatically declined in recent years because of poor policy decisions. Second, new competition policies will bring more broadband choices to small business consumers, driving market forces that lower prices and increase speeds to catch up with our global competitors. Finally, small businesses that depend on the Internet for e-commerce require policies like network neutrality that protect the free market, ensuring that there are no gatekeepers that obstruct their path to the market.

The problems we face today in the broadband access market are severe, but perhaps nowhere are they worse than in the small business sector. The problems in the residential market get the headlines and scrutiny. It is no secret that we are falling behind the world leaders in broadband penetration -- our broadband speeds are comparatively low and prices are high. Many small businesses (and particularly those with Internet-based goods and services) have a single choice for broadband service -- the incumbent telephone company. Compare that to global competitors in Europe and Asia that can choose from literally dozens of providers. The competitive market abroad translates into service that is far faster and less expensive. The economic disadvantages for our homegrown entrepreneurs over time are clear and the damage will be difficult to reverse.

Recent broadband policy at the Federal Communications Commission (FCC) has not embraced a free market approach to enabling competition, instead supporting the entrenched incumbency of telephone and cable companies. The legacy of these decisions has put downward pressure on investment opportunities and innovation in the small business sector. Right now, the FCC is considering a number of critically important regulatory choices -- including changes to the special access market and the barriers to market entry for competitive providers. Wrong decisions will result in higher broadband prices for small business and cripple competitive markets in ways that will take years to correct. In many cases, the incumbents seek to evade laws that foster competition through regulatory forbearance. Yet few in the Congress are paying close attention.

We recommend this Committee, working with the Small Business Administration (SBA), undertake a sweeping inquiry into the broadband policies that will directly benefit American small business. To begin, we need to improve our knowledge of the small business broadband market. Currently no federal agency is conducting serious data collection or analysis. We recommend the Committee support a variety of policy initiatives to bring competition to the marketplace including: ensuring spectrum auctions produce real competitors not vertical integration; opening the television white spaces for unlicensed use; protecting the rights of local governments to offer broadband services; guaranteeing the interconnection of networks on nondiscriminatory terms; transitioning Universal Service Fund (USF) programs to broadband; and safeguarding the Internet's free market for goods and services through network neutrality rules. We look forward to working with the Committee as it moves forward.

Broadband's Centrality to the Small Business Economy

It is now widely understood that the availability and adoption of broadband Internet access in our communities translates into jobs, investment and economic growth. For small business, it is an essential tool in the information economy -- a means to grow sales, expand to new markets, and innovate. Broadband is also rapidly becoming a difference-maker in a globally competitive market for goods and services. As the U.S. falls behind the world's leading nations in broadband penetration rates, speeds, and prices, the impact on entrepreneurs and small businesses will be severe. It is not merely that our counterparts in Europe and Asia have more broadband services to choose from -- they can often purchase ten times the speed at half the price. Using this technological edge, these companies can outperform U.S.-based competitors.

Broadband is not only important for keeping existing small businesses competitive; it is also critical in the creation of new small business jobs at home. A 2007 study by researchers at the Brookings Institution and MIT estimated that a one-digit increase in U.S. per-capita broadband penetration -- the metric used by the Organisation for Economic Co-Operation and Development (OECD) -- equates to an additional 300,000 jobs. [1] Thus our slide from 12th to 15th place in the world's broadband rankings during the latter half of 2006 equals approximately 240,000 lost jobs. [2] If our broadband penetration were as high as number-one-ranked Denmark, we could expect approximately 3.7 million additional U.S. jobs. This is not merely a matter of national pride; this is serious money and a life-or-death situation for the small business market. Small businesses often run on thin margins and innovative ideas, both areas that are squeezed if broadband technologies are unavailable or very expensive.

In 2005, the Small Business Administration (SBA) commissioned a study about broadband use by rural small businesses. [3] The study found: "Broadband investment and services appear to stimulate economic productivity and output, as well as create jobs." [4] The report summarizes a number of studies that confirm this finding and concludes that the conventional wisdom is correct. The primary finding in this report is that rural small businesses are less likely to have broadband services and more likely to miss out on the economic benefits broadband brings. The report does not make any international comparisons to note the competitive disparity between the U.S. and international markets. However, it does note that communities with broadband services "have a competitive edge in terms of attracting and retaining businesses" [5] -- a critical component of economic development. This finding is applied to different U.S. towns and cities, but it is equally true of a comparison between the U.S. and Europe or Asia.

Increasingly, good business depends on good communications technologies. Manufacturers increasingly require online inventory and ordering capabilities for sales points. According to the Census Bureau, 92 percent of e-commerce takes place business-to-business. These transactions rely "overwhelming on proprietary Electronic Data Interchange (EDI) systems." [6] Small businesses

---

[1] Robert Crandall, William Lehr and Robert Litan, "The Effects of Broadband Deployment on Output and Employment: A Cross-sectional Analysis of U.S.Data, June 2007. Available at http://www.brookings.edu/views/papers/crandall/200706litan.htm.

[2] Organization for Economic Cooperation and Development, "OECD Broadband Statistics to December 2006", http://www.oecd.org/sti/ict/broadband.

[3] Stephen B. Pociask, "Broadband Use by Rural Small Business," December 2005, Small Business Administration, Office of Advocacy, Available at http://www.sba.gov/advo/research/rs269tot.pdf

[4] Ibid, i.

[5] Ibid, 3.

[6] "E-Stats," US Census Bureau, May 25, 2007, Available at http://www.census.gov/eos/www/2005/2005reportfinal.pdf (Figures are for 2005, the last year reviewed in this study).

without the communications capacity necessary to take advantage of EDI systems are left out of this multibillion-dollar industry. In addition, advertising and marketing are increasingly done online and the web interface for a small business is often as critical as its brick-and-mortar façade. Many small businesses -- such as the thousands of eBay power-sellers -- are exclusively online. Retail e-commerce sales totaled $33.6 billion in the second quarter of 2007, up 20.8 percent from the same quarter in 2006.[7]

Certainly, the primary interest for this Committee in broadband policy must be to increase the number of broadband choices in the small business market in order to increase speeds and lower prices. But it is important to note that small businesses are not just the beneficiaries of better broadband services. Competitive broadband providers, i.e. new entrant Internet Service Providers (ISP) and Competitive Local Exchange Carriers (CLEC), are often small businesses themselves (assuming a definition of small business as having less than 500 employees). These businesses -- as well as those that use the networks to transact commerce -- rely on a free market for the production, consumption and transmission of Internet packets. This is the reason why small businesses have been central to the network neutrality debate raging for almost two years. Any imposition of gatekeepers in the access market will jeopardize the engine of innovation in the small business economy.

Broadband Market Failures in the Small Business Sector

The small business broadband marketplace is in a state of alarming failure. Not least of our problems is the fact that no government agency monitors the small business broadband market. We must extrapolate the state of the market by making informed assumptions about the residential broadband market (from which most small businesses buy their services) and the enterprise market for broadband. The FCC collects no data specific to small business broadband connections. The SBA's 2005 study laments this fact and calls for more research and better measurements. Virtually nothing has been done to address this glaring lack of data. We cannot fix problems that we do not measure.

The SBA did conduct a survey in 2004 to determine whether or not small businesses are subscribing to broadband, what type of service they buy, and what price they are paying.[8] Although three years is a long time in the broadband market, a number of findings are worth noting because they reveal very significant problems which almost certainly have not been remedied.

Extrapolating from the SBA survey data, the marketplace for small business broadband connections resembles the residential broadband market because the vast majority of small businesses are buying consumer-class connections (i.e. asymmetric upload/download speeds without a dedicated line). These asymmetric lines are sometimes marketed as "business-class", but they do not have the reliability of a dedicated line or the functionality of symmetric upload speeds. Broadband lines that are not dedicated to one customer are often shared by 20 to 50 customers -- a metric known as the contention ratio. These figures are proprietary to each broadband provider and are not made public -- so we cannot know for sure what kinds of services are actually in the market. As a technological matter, there are no cable modem products that are symmetrical in speed and offer dedicated lines. Only a T-1 service or better in digital subscriber line (DSL) or fiber-optics can

---

[7] "Quarterly Retail E-Commerce Sales, 2nd Quarter 2007" US Census Bureau, August 16, 2007, Available at: http://www.census.gov/mrts/www/data/pdf/07Q2.pdf

[8] "A Survey of Small Businesses' Telecommunications Use and Spending," Stephen Pociask, TeleNomic Research for the Office of Advocacy, Small Business Administration Contract No. SBA-HQ-02-M-0493, Washington, DC, March 2004 available at http://www.sba.gov/advo/research/rs236tot.pdf

provide this level of service. The standard T-1 line -- typically a 1.5 Mbps symmetrical connection with capacity dedicated to one end-user business customer -- often prices small businesses out of the market, even though its speed is hardly revolutionary.

According to the SBA survey, only 4 percent of small businesses were buying T-1 lines in 2004. Even if we generously assume that this number has tripled in the last three years, this is a huge problem by itself. It indicates that most small businesses do not subscribe (because of price or availability) to the kinds of communications technologies that are best suited to business use. These business-class broadband capabilities are available at higher speeds and much lower prices in international markets -- which points to a glaring competitive disadvantage at home. If an IT consulting firm in Massachusetts is serving its clients from servers connected to a cable modem (8 Mbps download/3 Mbps upload), and its Japanese counterparts are competing for the same clients with servers connected to a fiber-optic line (100 Mbps upload and download), the situation is not sustainable for U.S. business interests. Adding insult to injury, the Japanese firm likely pays the same price or less for its connection!

For any business that pushes data out from its own servers, the most important problem is not download speed, it is upload speed. For a small business that deals in e-commerce, markets products online, provides services or processes orders over the network, or communicates between offices via a high-speed line, it is critical to have sufficient upload speed to transmit data to clients and consumers. Reliability is also a critical factor. If the business depends on the network connection, it cannot go down. If these are the main concerns for small businesses, any small business without at least a T-1 line will be at a competitive disadvantage. The SBA data shows how far the marketplace is to realizing universal adoption of these kinds of services.

The central problem is that there is insufficient competition in the marketplace for T-1 class connections to lower prices to a reasonable level. Because cable operators do not offer these services, the incumbent telephone company has a monopoly -- unless there are CLECs in a particular local market. Moreover, prices fluctuate dramatically across the country, leaving rural areas at a tremendous disadvantage.

In a recent workshop[9] hosted by the Cooperative Association for Internet Data Analysis (CAIDA), practitioners reported the prices they pay for 1 symmetrical Mbps of dedicated broadband service in different areas of the country. There are no publicly available datasets that provide this kind of price information on a larger scale, but this snapshot gives a striking display of the disparities and the importance of supply-side market conditions. There is an urgent need both to study this problem and use policy changes to mitigate the worst of the damage.

> **Price per month of 1 symmetric Mbps of dedicated broadband service:**
>
> San Francisco -- $8-12
> Chicago, Ill. -- $80-90
> Urbana, Ill. -- $300-320
> Greenup, Ill. -- $1300

These figures are corroborated by the SBA's survey that reports the average monthly expense on a T-1 line for small businesses at $720.[10] There are multiple factors at play in the price

---

[9] Commons Project Strategy Summit, December 2006, San Diego, CA, Cooperative Association for Internet Data Analysis (CAIDA).

[10] See summary of SBA survey: Podiask, 19.

disparity here (including the cost of service provision in rural versus urban areas), but by far the most important one is the presence of competitive service providers. As a general rule of thumb, the more CLECs there are in a market, the more likely it is that prices are driven down.

The consequences of this problem are stark. By way of analogy, imagine if small businesses faced similar disparities in gas prices. By analogy, if a gallon of gas were $2 in San Francisco, it would cost $260 a gallon in Greenup. Is there any wonder where investment, jobs and economic growth will go in such an environment?

This is where the rural digital divide and the international comparisons become very significant for the small business economy. If the communications technologies most appropriate for business users are unavailable or excessively priced in rural areas, those businesses will either never materialize, or they will move to urban areas. A 2005 survey reported that three-quarters of rural small businesses did not have access to the broadband technologies they need.[11] If these technologies are available at higher speeds and lower prices overseas than anywhere in the U.S. market (rural or urban), either the jobs will flow abroad or the competitive advantages will tip the scales dramatically against the U.S. economy. The SBA survey reported that the average small business customer that did not have a T-1 line paid between $40 and $50 per month for asymmetric cable modem and DSL service. These connections are typically between 3 and 8 Mbps on the download and roughly one-third of that or less on the upload. By contrast, connections in France are 3 to 10 times those speeds for the same price.[12] In Japan, the same money buys 8 to 30 times that speed.[13]

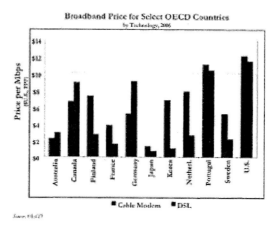

Broadband Price for Select OECD Countries
by Technology, 2006

Source: OECD

[11] Gross, Grant, "Survey: Small businesses lack broadband options," IDG News, September 20, 2005, Available at http://www.infoworld.com/article/05/09/20/HNsmbbroadband_1.html
[12] See: Jennifer L. Schenker, "Vive la High Speed Internet," Business Week, July 18, 2007, Available at: http://www.businessweek.com/print/globalbiz/content/jul2007/gb20070718_387052.htm
[13] See: Blaine Harden, "Japan's Warp-Speed Ride to Internet Future," Washington Post, August 29, 2007, Available at: http://www.washingtonpost.com/wp-dyn/content/article/2007/08/28/AR2007082801990_pf.html

A recent comprehensive survey by the OECD indicates that the U.S. small businesses that choose to purchase the more expensive but more reliable symmetrical leased-access connections pay far more than business users in most other OECD nations.[14] The OECD found that while businesses in countries like Denmark and Iceland pay approximately $350 USD per month for a 2 Mbps leased-access line, U.S. businesses are paying on average $2,500 per month for the same product.

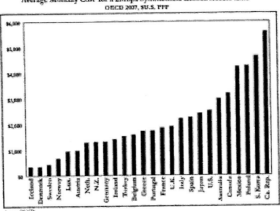

Average Monthly Cost for a 2Mbps Symmetrical Leased Access Line
OECD 2007, $U.S. PPP

It is imperative that Congress pay close attention to broadband regulatory policy issues -- great and small -- in order to ensure that by the time Capitol Hill moves to address the disastrous failures in the broadband marketplace, the FCC hasn't already given up the game. As this analysis demonstrates, the key problem is not the availability of broadband services (although that is a significant problem in many rural areas); it is the creation of competitive markets to make them faster and more affordable. How do we get T-1 class or better services in as many markets as possible? How do we open the market to more providers to create the competition which will drive the costs down so that they are affordable to most small businesses?

Notwithstanding the absence of specific data and analysis on the small business broadband market, Congress is acutely aware of the larger problems in the marketplace. The facts are unambiguous in the residential broadband market (which we have seen has a large overlap with the small business market). A significant number of American households -- around 10 percent -- have no available terrestrial broadband service.[15] A much larger percentage -- over 40 percent -- have service available to them, but they do not subscribe -- foregoing the social and economic benefits of connectivity because of high prices, a lack of equipment and training, or simple disinterest.[16] Rural

---

[14] "OECD Communications Outlook 2007", Organization for Economic Cooperation and Development, Information and Communications Technology Division, July 2007.

[15] "Broadband Deployment is Extensive throughout the United States, but it is Difficult to Assess the Extent of Deployment Gaps in Rural Areas," Government Accountability Office, Report to Congressional Committees, GAO-06-426, May 2006.

[16] Extrapolated from "High-Speed Services for Internet Access as of June 30, 2006," Industry Analysis and Technology

areas lag behind urban areas in broadband access.[17] These inequalities present significant downward pressure on the likelihood that small businesses will be founded and succeed in the geographic and socio-economic areas most in need of economic development.

The cost to our economy and the quality of life in our society mounts each successive year that these problems go unsolved. Meanwhile, alarmingly, the U.S. is falling behind the rest of the world in broadband penetration and market performance, ceding the tremendous benefits of leading the world in network connectivity to others. The President called for us to reach the universal broadband milestone by this year. There is now no chance we can achieve that result. While it is true that the total number of broadband lines deployed in the U.S. is rising and the total number of broadband users is now near 50 percent of the country, the U.S. growth rate in broadband penetration compared to other nations is not encouraging. Our growth rate between 2005 and 2006 earned us the 21st spot out of 30 among OECD nations.[18]

Country	Year to Year Absolute Change In Broadband Penetration (OECD)					
	Dec 2001-2002	Dec 2002-2003	Dec 2003-2004	Dec 2004-2005	Dec 2005-2006	Slowing ('05 to '06)?
Denmark	3.8	4.8	6.4	5.9	7.0	No
Netherlands	3.2	4.8	7.2	6.2	6.6	No
New Zealand	0.9	1.0	2.1	3.4	5.9	No
Ireland	0.3	0.5	2.5	3.4	5.8	No
Sweden	2.7	2.6	3.8	5.7	5.8	No
Norway	2.3	3.8	6.8	7.0	5.7	Yes
Hungary	0.3	1.4	1.6	2.7	5.6	No
Luxembourg	1.2	2.0	6.3	5.1	5.5	No
Australia	0.9	1.7	0.2	6.1	5.4	Yes
France	1.8	3.1	4.6	4.6	5.2	No
United Kingdom	1.7	5.1	5.1	5.9	5.2	Yes
Finland	4.2	4.0	3.4	7.5	4.8	Yes
Poland	0.2	0.5	1.3	0.3	4.5	No
Switzerland	3.6	4.5	7.4	6.6	4.4	Yes
Belgium	4.3	3.0	3.8	2.7	4.3	No
Czech Republic	0.1	0.3	2.0	3.9	4.2	No
Germany	1.8	1.5	2.8	4.6	4.1	Yes
S. Korea	4.6	2.4	0.6	0.4	3.9	No
Spain	1.8	2.4	2.7	3.4	3.8	No
Iceland	4.7	5.9	3.9	8.2	3.3	Yes
United States	2.4	2.8	3.2	5.4	3.3	Yes
Greece	0.0	0.1	0.3	1.9	3.2	No
Slovak Republic	0.0	0.3	0.7	1.5	3.2	No
Austria	2.0	2.0	2.5	6.2	3.0	Yes
Italy	1.0	2.4	4.0	3.7	3.0	Yes
Canada	3.2	3.0	2.5	3.4	2.8	Yes
Japan	3.9	4.6	4.3	2.6	2.6	Yes
Portugal	1.5	2.3	3.4	3.3	2.3	Yes
Turkey	0.0	0.3	0.4	1.4	1.7	No
Mexico	0.2	0.1	0.5	1.3	1.3	Yes

Despite the inactivity of the agencies responsible for broadband deployment, the broadband problem is well-documented. Accordingly to the best available data:

Division, Wireline Competition Bureau, Federal Communications Commission.; calculated assuming one line per household, based on July 1 2006 Census household estimates; S. Derek Turner, "Broadband Reality Check II," Free Press, Consumers Union, and Consumer Federation of America, August 2006, Available at http://www.freepress.net/docs/bbrc2_final.pdf

[17] SBA's study affirms this finding. See Pociask, op cit.

[18] Organization for Economic Cooperation and Development, "OECD Broadband Statistics to December 2006", http://www.oecd.org/sti/ict/broadband.

- **Extrapolating from FCC data, nearly 60 percent of U.S. homes are not broadband adopters.**[19]
- **The rate of residential broadband adoption continues to slow.** From June 2005 to June 2006 the number of residential advance service lines increased 34 percent. But from June 2004 to June 2005 the increase was 62 percent.[20]
- **37 percent of ZIP codes have one or less cable and/or DSL provider.**[21] Given that FCC ZIP code data overstates the level of broadband deployment; this should be viewed as a conservative figure.
- **Some states have large gaps in coverage.** Over 40 percent of South Dakota households are not wired for cable broadband. Over 40 percent of New Hampshire and Vermont households are not wired for DSL.[22]
- **The broadband market remains a duopoly.** 96 percent of residential advanced-services lines are either cable or DSL.[23]
- **There are no viable 3rd "pipe" competitors.**

  o From June 2005 to June 2006 there were only 637 new broadband over powerline (BPL) connections added, bringing the total to just over 5000 nationwide, or 0.008 percent of all U.S. broadband connections.[24]

  o From December 2005 to June 2006 the number of advanced service satellite broadband connections DECREASED by 40 percent.[25]

  o Mobile wireless broadband from cellular carriers enjoyed a rapid growth-rate in the last year. However, these connections remain slow and costly compared to wireline alternatives. They are not substitutable competitors with DSL and cable modem, but rather form a complementary market dominated by vertically integrated firms with little incentive to cannibalize wireline market share.

  o The likelihood of solving the small business broadband problem with a wireless third pipe is even more remote than the notion that residential wireline services will be replaced with wireless. The needs of small business for higher speeds and symmetrical connections at affordable rates stand in direct contrast with the characteristics of asymmetrical, slow, expensive wireless connectivity

This record of performance has not positioned us well in the race for global competitiveness -- with all of the economic and social benefits at stake. According to the OECD, the U.S. is 15th among the 30 member nations in broadband penetration, lagging behind the acknowledged world

---

[19] "High-Speed Services for Internet Access as of June 30, 2006," Industry Analysis and Technology Division, Wireline Competition Bureau, Federal Communications Commission; calculated assuming one line per household, based on July 1 2006 Census household estimates.

[20] Ibid

[21] Ibid

[22] Ibid

[23] Ibid

[24] Ibid

[25] Ibid

leaders, the Netherlands and South Korea, but also Canada and all of Scandinavia.[26] The International Telecommunication Union (ITU), evaluating a larger number of countries than the OECD, places the U.S. at 16th.[27] A separate ITU study measuring a variety of factors in the Digital Opportunity Index, places the U.S. at 21".[28] The consequences of lagging performance are severe.

Current Policy Debates Affecting Small Business Broadband Market

Buried in the arcane world of telecommunications regulatory policy are a number of issues that have enormous bearing on the quality, price and competitive availability of business-class broadband connections. For example, the FCC is currently mired in debates over three technical regulatory proceedings: special access regulation, forbearance petitions on unbundling and interconnection requirements, and the retirement of copper wire facilities by incumbent telephone companies. What does this mean for small business?

The special access debate centers on the prices that incumbent network owners charge to competitive service providers to transport and terminate the aggregated traffic from the competitive last-mile networks. Incumbent Local Exchange Carriers (ILEC) are pushing to deregulate pricing of their monopoly infrastructure. CLECs, wireless telecommunications providers (e.g. Sprint/Nextel, T-Mobile), cable companies and municipal broadband providers all pay special access rates to the incumbent networks (usually Bell companies) that own the backbone of the Internet and the regional networks that carry traffic to the backbone. These competitors are arguing that prices should continue at rates that permit competition. Generally speaking, the higher the rates are for special access, the higher the prices are for consumers of telecommunications services from these kinds of competitive service providers, since the costs the competitors pay to the incumbents must be passed along to consumers. Since CLECs disproportionately serve the business market, this debate is highly significant for the future of small business connectivity. Special access rates will play a huge role in determining the cost and availability of T-1 or better classes of business broadband service.

The market is hardly unprofitable for the incumbents. According to Sprint/Nextel's testimony before the Senate Commerce Committee in 2006: "Just last year ATT/SBC earned a rate of return of 92 percent on its special access services; BellSouth earned nearly 98 percent."[29] In 2005-2006, the special access market was a $16 billion business. Over 80 percent of this revenue went to Verizon and AT&T. The profit margin in this sector was between 50 and 100 percent. These monopoly rents stand in stark contrast next to the FCC's authorized rate for rate-of-return carriers, 11.25 percent. Clearly, there is little competition in special access.[30] The Government Accountability Office (GAO) confirms this finding in the marketplace, reporting on its study of telecommunications companies providing service to business districts: "Data on the presence of competitors in commercial buildings suggest that competitors are serving, on average, less than 6 percent of the buildings with demand for dedicated access in these areas. For buildings with higher

[26] Organization for Economic Cooperation and Development (OECD), "OECD Broadband Statistics to June 2006," October 13, 2006, Available at: http://www.oecd.org/sti/ict/broadband

[27] http://www.itu.int/ITU-D/ict/statistics/at_glance/top20_broad_2005.html

[28] World Information Society Report, August 2006, http://www.itu.int/osg/spu/publications/worldinformationsociety/2006/wisr-web.pdf

[29] Testimony of Robert S. Fonsaner, Senior Vice President – Government Affairs, Sprint Nextel Corporation before U.S. Senate Committee on Commerce, Science and Transportation, June 13, 2006.

[30] Sprint Nextel Corporation Comments to the Federal Communication Commission, May 16, 2007, Docket No. 07-45.

levels of demand, facilities-based competition is more moderate, with 15 to 25 percent of buildings showing competitive alternatives, depending on the level of demand."[31]

In addition, the incumbent providers have filed numerous forbearance petitions at the FCC regarding various regulations that, among other things, control special access rates and require wholesaling of network elements to CLECs. They seek forbearance -- meaning they are requesting the FCC simply decline to enforce the rules that govern them -- in order to undo the regulations that create competitors in their markets. Business class services are the primary arena of dispute, once again casting a direct line of influence on small business customers. If the incumbents are granted forbearance from the rules, the rates for small businesses seeking first class broadband service could increase very substantially. The competitive pressures that have exerted what little pressure exists on pricing will be gone.

Finally, there is a hot debate over what is known as "copper retirement." According to numerous press reports and the complaints of the CLECs, incumbent telephone companies (notably Verizon) are decommissioning or even cutting the copper wire when they install new fiber-optic lines into a neighborhood.[32] The result is that the CLECs that were paying Verizon to use those lines are now unable to compete in that market. They cannot reach their customers! Verizon claims that it is unacceptable to ask them to run two networks -- a fiber and a copper network. However, in the likely event competitors will run the networks, they need to buy or lease them intact, not inert and useless in the ground. Beyond the importance of maintaining competitive markets, these wires should not be Verizon's to retire. They have been paid for many times over by the rate paying public. They have also been fully depreciated through tax incentives for the Bell companies. And, of course, they are laid on the public's rights-of-way. Once again, the competitive service providers that are losing out in this debate are the industry that specializes in business class broadband services.

These issues carry a great deal of importance for the future of small business broadband competition. They are often considered independent of the larger focus of the Congress on pro-competitive broadband policy and the goals of increasing speeds and lowering costs by triggering market forces. But they are tied to that mission. If each of these debates results in the reduction of competition, they will weigh down and inhibit the progress toward a better broadband marketplace for small business, further reducing global competitiveness across the economy.

Policy Agenda to Address the Broadband Problem for Small Business

Clearly, there is a strong need to address our growing broadband problems. Perhaps nowhere is the urgency more pressing than in the small business marketplace. Most of our small businesses are not buying the services best suited to them because of cost, even as their global competitors race ahead. Even if we correct course immediately, it will take years to undo the damage.

The first step is establishing a serious national broadband policy. Currently, we are "the only industrialized state without an explicit national policy for promoting broadband."[33] According to FCC Commissioner Michael Copps: "We recently got a commitment on a goal, on an objective. But

---

[31] See findings: United States Government Accountability Office, "FCC Needs to Improve Its Ability to Monitor and Determine the Extent of Competition in Dedicated Access Services," GAO-07-80, November 2006.

[32] See for example: Ed Gubbins, "CLECs protest copper retirement," *Telephony Online*, May 21, 2007, Available at: http://telephonyonline.com/mag/telecom_clecs_protest_copper/

[33] Thomas Bleha. "Down to the Wire." *Foreign Affairs*, May/June 2005. http://www.foreignaffairs.org/20050501faessay84311/thomas-bleha/down-to-the-wire.html

an objective and a strategy are two vastly dissimilar things."[34] The key problem is that U.S. broadband policies have not engaged free market competition, choosing instead to deregulate incumbents and wait for the elusive intermodal competition of wireless and BPL to come along and challenge the stagnant duopoly of DSL and cable. This policy will not work for the residential market -- where redundant infrastructures have brought complimentary, not substitute, broadband services. This policy *cannot* work for the business market, where the most suitable services are only available on a single network. Small businesses that buy either consumer or business class broadband will rise and fall in the global marketplace based on the number of choices they have for broadband and the price per unit of speed.

We need to identify our goals for the small business broadband market and work backward to find the right policies. We suggest goals that address our shortfalls in each of the three major indices of broadband performance: availability, price and value (cost per unit of speed).

Goal #1 -- Establish universal availability of business-class broadband services

Goal #2 -- Lower barriers to market entry for competitive ISPs -- stimulating market forces to drive prices down and speeds up

Goal #3 -- Stabilize the market conditions that will permit small businesses to move out of the consumer-grade broadband market and subscribe to affordable, business-class services.

To regain global leadership in broadband and maximize the social benefits of a network economy, we need to establish a framework that supports an evolving communications infrastructure that will ultimately provide 100 Mbps of symmetrical connectivity to small business in America in the next decade. This is the standard that has already been reached by the world's leading broadband nations. We have no time to lose.

To achieve the goal, we will need vigorous, multimodal competition -- that is, competition between delivery platforms (e.g. DSL, cable, and wireless) as well as competition within delivery platforms (e.g. multiple ISPs offering T-1 service in a market). We cannot and should not bet our digital future on one form of competition. These competition policies will provide healthier markets for small business consumers of broadband as well as prompt the emergence of small business ISPs carving out sectors of the market for their own innovative offerings.

We should also ensure that the content/applications market that sits adjacent to the connectivity/access market also retains maximum competitiveness. Through network neutrality rules, we can preclude market power in network ownership from distorting the market for Internet content. This will maximize innovation among small businesses in the content and services market, stimulating greater investment and job growth in the sector. It will also ensure that small businesses compete on a level playing field with large businesses. To realize these goals, we will need to establish a national broadband policy framework that is comprehensive and aggressive in pursuit of market competition and advanced network capabilities.

Study the Problem

We should begin by addressing our data problems. This Committee should press the SBA to conduct further studies in conjunction with the FCC on the small business market. There is no

---

[34] Jim Hu, "Why Our Broadband Policy's Still a Mess," CNet, February 28, 2005, Available at: http://www.news.com/Why-our-broadband-policys-still-a-mess/2008-1034_3-5590929.html

specific information at any federal agency on small business connections in the U.S., which inhibits our ability to craft the right policies. We should also study the international competitiveness of our small businesses, focusing on ways we can bridge the technology gap to the world leaders.

We should study the cost and feasibility of broadband technologies. We do not have reliable cost estimates for deploying different technologies to meet the needs of business broadband users. For years, we have heard that technologies like BPL and satellite wireless broadband were inches from transforming the marketplace. Yet we did not study these issues sufficiently to determine that those estimates were overblown and unrealistic. A paucity of information has led us to false expectations and delay, distracting from the need to seek out the necessary data points to make policy.

To do all of this effectively, we needed better data in general. We need to know at a granular level – block by block – where broadband service is available and where it is not. But we must go beyond that. We must collect information about the price and speed of connections as well. Without this information, we cannot quickly identify the gaps in the service market and remedy market failures that hold prices high and service quality low.

Programs like ConnectKentucky represent a valuable model to consider for federal policy— particularly in its focus on working with local communities. The ConnectKentucky model has much in it to recommend. In particular, the combination of teams of local stakeholders with localized broadband data collection is a useful method to aggregate market demand and attract the cooperation of broadband carriers. This brand of on-the-ground needs assessment is a very useful innovation in the sector—though it does raise perplexing questions about the quality of the carriers' own market research.

However, there are limitations with the ConnectKentucky model. The data the program collects is exclusively proprietary. This means that the information about deployment in different geographic areas cannot be used by researchers, business leaders and policymakers to further inform policy and investment decisions. Further, the program does not collect information about price and speed of broadband connections. This is a significant limitation. It is particularly problematic in areas which are not wholly unserved but nonetheless have low broadband penetration rates. Finally, if programs like ConnectKentucky were to be instituted nationwide on a state by state basis, the information collected that can be made public would not be comparable between states and the insights from a bigger picture analysis would be unavailable.

Enact Multi-Modal Competition Policy

The problems in the marketplace will not be solved by tweaking around the edges; nor will they be solved by enacting policies that are subsidies of status-quo, incumbent business models. We need to reject the conventional political wisdom of complacent incrementalism and embrace a policy inquiry into all the possible options for putting our broadband future back on track. Now is not the time to make artificial declarations that some ideas are off the table and narrowly focus on particular proposals. No one policy idea is the silver bullet. It will require many different initiatives aimed at different levels of the broadband market to accomplish our goals. In short, it must be "multimodal" – by which we mean that it must foster competition both *within* and *between* broadband technology markets.

We present here an outline that may serve as a blueprint of ideas for a national broadband policy that serves the interests of small business. We would encourage other stakeholders to offer the Committee similar, comprehensive proposals for consideration. To simplify for present purposes, the broadband market can be understood as two separate arenas: 1) a physical connection to the Internet and the technologies used to transmit information over the network; and 2) the applications and content delivered via that Internet connection and the devices used to receive them. We can and should target broadband policy in both layers of the network to maximize the productivity of both markets.

Policies for the Physical Layer

Given the dearth of small businesses that subscribe to symmetrical, dedicated broadband connections, the first policy priority must be expanding the reach, capacity, competitiveness and efficiency of our networks to serve small business customers. In turn, these networks support the spread of advanced Internet applications that carry the nation's growing e-commerce business.

➢ Reasonable and nondiscriminatory interconnection between facilities-based providers — Since the Internet is nothing more than a global network of interconnected private and public networks, it is imperative that each interconnects with one another to maximize the efficiency and utility of the overall network. This policy is central to the revitalization of a competitive marketplace for business-class broadband services.

➢ Reintroduce intramodal competition into the broadband market — Though recent FCC decisions have moved away from this model of competition policy, it is imperative that it is not abolished. Intramodal competition through open access to network infrastructure has been the cornerstone of international broadband successes. Forbearance petitions seeking to circumvent these rules at the FCC should be denied.

➢ Pro-competitive regulations for special access telecommunications connections — The incumbent networks must not be permitted to price all competition out of the market and destroy what little remains of the competitive ISP industry. Forbearance petitions at the FCC seeking to circumvent undermine competition in the special access market should be denied.

➢ Allocation of licensed public spectrum aimed at creating wireless broadband competitors that are independent of wireline incumbents and offer capacity on a wholesale basis.

➢ Expansion of unlicensed public spectrum into lower frequencies by opening up the unassigned television channels (also known as "white spaces") for wireless broadband. We recommend the Kerry-Smith bill, S. 234.

➢ Reform and transition the federal universal service programs from dial-tone to broadband — We should move our valuable Universal Service Fund (USF) programs into the 21st century with targeted subsidies and accountability benchmarks to support broadband deployment in high-cost areas.

➢ Explore financial incentives to expand broadband capacity in the last mile — Successful policies overseas have included direct government investment in wiring public facilities, low-interest loans for public and private broadband projects, tax incentives for networking equipment, accelerated depreciation, debt guarantees and other targeted investments in our digital future.[35]

---

[35] Gross, op. cit.

> Authorize and protect the right of local governments to provide broadband services --
> Municipalities have led the charge in recent years to fill gaps in the broadband market and build
> services that exceed those offered by commercial incumbents. This effort to bring competition
> and innovation to the marketplace should be encouraged. We recommend the Lautenberg-
> McCain bill, S. 1853.

> Collect data and map the broadband market on an ongoing basis -- We cannot solve problems
> that we do not understand. Our current state of broadband data collection is unacceptable.
> FCC should be instructed to collect more granular information on service as well as price and
> speed data on all broadband connections. Programs should be initiated to specifically study the
> small business market. We recommend the Inouye data collection bill, S. 1492, which has
> recently passed out of the Senate Commerce Committee.

<u>Polices for the Applications Layer</u>

The applications layer, in this analysis, refers to the marketplace for content, applications, services
and devices that flow over, or connect to, the Internet. This economic space at the "edge" of the
network architecture has been a remarkable engine of economic growth in the small business sector
in the last decade. Innovators and entrepreneurs should have not barriers to entry to sell their ideas.
We need an absolutely free market, absent any gatekeepers. Policies aimed at the application layer
should recognize its centrality to the economic and democratic health of the nation.

> Network Neutrality should be established as the cornerstone of broadband policy -- We should
> protect an open market for speech and commerce on the Internet for consumers, citizens and
> businesses alike. To do this, we should apply nondiscrimination safeguards to the broadband
> ramps leading onto the Internet that prohibit owners of the physical layer of the network from
> gate-keeping the applications layer of the network. We recommend the Dorgan-Snowe bill, S.
> 215.

> *Carterphone* rules should apply to the wireless broadband platform -- We should recognize and
> remedy the contradictions in fostering an open market for wireless broadband on a platform
> emerging from the closed networks of cellular telephony. The walled garden of the personal
> communications service (PCS) world should not be permitted to cripple the potential of mobile
> wireless broadband. All devices, applications and services that do not harm the network should
> be permitted access.

> Facilitate ongoing research into network traffic and data management -- The dearth of
> information about what is happening on the Internet cripples our efforts to address some of the
> most pressing problems in the application layer: spam, cyber-security, privacy and traffic
> management. Policymakers should seek to make available the tools researchers need to provide
> the best available answers to these problems.

Conclusion

The broadband problems in the U.S -- and the small business broadband problem in particular -- are
urgently in need of redress. If we watch and wait, trusting that today's artificially constrained
marketplace will magically solve market failures, we will see the U.S. slip farther behind the rest of

the world and widen the digital divide -- both domestically and internationally. The consequences are too severe to permit. Even if we reversed engines today, it would take years to catch up to the world's leading broadband nations.

The way forward is clear, it simply requires the political will to recognize the problem and address it with swift and comprehensive policy change. Broadband is now well understood to be a driver of economic growth and an essential part of a healthy small business sector. Yet the lack of competition in the broadband market is so severe that most small businesses are unable to purchase the kind of broadband service most suited to advance their competitive interests. Many small businesses -- especially in rural areas -- do not have connectivity at all. Meanwhile, the gap to our global competitors is widening across the board. The losses we are incurring as a result of the status quo are measured in billions of dollars.

In spite of these harsh realities, we still lack a comprehensive national broadband policy. If anything, our current policies are headed in the wrong direction. The incumbent network owners are busy pressuring the FCC to permit them to sweep away the last free market policies on the books and crush what little competition remains. If they are successful, the only market forces exerting downward pressure on the prices for business class broadband service will disappear. As global broadband markets are flooding with competitive offerings, ours are contracting.

Perversely, the proposals of the incumbents also include dismantling the open, neutral marketplace for commercial applications to squeeze out higher revenues at the expense of new innovators. The result in the value chain will be a resounding net loss. This is robbing Peter to pay Paul. We must reject the argument that an open Internet and a high capacity network are mutually exclusive goals. We must have both for our information marketplace to prosper. Nowhere is this truer than for American small business.

The first step on the road to broadband recovery is understanding the problem. We must rectify the deplorable state of data collection in the broadband market. What we do not know undercuts our ability to craft and target viable solutions. Armed with the right information, the Congress should move forward with a comprehensive national broadband policy. This should be a broad platform of initiatives that addresses the complexity of the issue and maximizes our chances for near and long term success. The focus of these policies should be: 1) enhancing competition between and within the technologies that deliver broadband connectivity; 2) protecting competition and speech in the content flowing over the Internet; 3) expanding opportunities to bring new broadband providers to the market using new technologies; 4) using targeted economic incentives to stimulate investment in underserved areas; 5) promoting a permanent research agenda that facilitates the collection of data in the market and on the network. We look forward to working with the Committee to support these productive goals.

---

[1] Free Press is a national, nonpartisan organization with over 350,000 members working to increase informed public participation in media and communications policy debates.

[2] Consumers Union is a nonprofit membership organization chartered in 1936 under the laws of the state of New York to provide consumers with information, education and counsel about goods, services, health and personal finance, and to initiate and cooperate with individual and group efforts to maintain and enhance the quality of life for consumers. Consumers Union's income is solely derived from the sale of *Consumer Reports*, its other publications and from noncommercial contributions, grants and fees. In addition to reports on Consumers Union's own product testing, *Consumer Reports* with more than 5 million paid circulation, regularly carries articles on health, product safety, marketplace economics and legislative, judicial and *regulatory* actions which affect consumer welfare. Consumers Union's publications carry no advertising and receive no commercial support.

[3] The Consumer Federation of America is the nation's largest consumer advocacy group, composed of over 280 state and local affiliates representing consumer, senior, citizen, low-income, labor, farm, public power and cooperative organizations, with more than 50 million individual members.

Chairman KERRY. Ben, thank you very much. A quick comment: I posted a blog this morning on Free Press and there were very thoughtful responses. I think there are about 72 at this moment. I am going to put this in the record, the responses that came in, and Senator Snowe, I will get a copy to you, but they are really thoughtful with a lot of folks raising questions about whether or not you should treat this as a public utility, all of them appalled by the lack of competition, the lack of access, suggesting ways in which we might be able to get it. So thank you for the testimony. It is very important and we appreciate it.

[Response to Senator Kerry's blog appears in the appendix on page 137.]

Mr. Mefford, welcome.

STATEMENT OF BRIAN MEFFORD, PRESIDENT AND CHIEF EXECUTIVE OFFICER, CONNECTED NATION, BOWLING GREEN, KENTUCKY

Mr. MEFFORD. Chairman Kerry, Ranking Member Snowe, thank you for the opportunity to be with you today. I appreciate the invitation.

I want to begin my testimony with a bit of a story that represents what we are seeing in Kentucky and what the types of opportunities are that are all about us as Kentucky has moved close to ubiquitous broadband coverage. It is the story of an entrepreneur named Kamren Colson who grew up in the "Burley Belt"of Central Kentucky, and like too many Kentuckians, after he graduated from college, couldn't find opportunities near home, and moved to a place that was more conducive to the creative class.

He began a graphic design company and operated that company for a few years and then decided around 2004 or 2005 that he was going to push the broadband envelope—this whole technology opportunity—and so he said, I have this family farm that I grew up on in Kentucky and we don't raise tobacco anymore and it is just kind of sitting there. And so he said, I am going to relocate my business to Central Kentucky. And he said, with broadband technology, I can connect to my potential clients—my clients—just as easy as I can from a downtown business center.

And so he did that. About a year after moving to Kentucky, he and his business won the account for creating the 2006 Academy Awards program and all the additional promotional assets for the Academy Awards. So from a former tobacco field in Central Kentucky, this creative design services firm was operating back and forth with folks in Los Angeles as if they were down the hall from the

Academy. The Academy reported that it was no different. They said they didn't even realize that he was in another State and it was just like he was down the hall.

That is not an isolated example, but rather an illustration of what is happening throughout Kentucky as we move closer to 100- percent broadband coverage. And I will tell you that based on the broadband that was deployed in 2005 alone, Kentucky has saved or created 59,000 jobs. In the technology sector, in the last 2 1/2 years, Kentucky has created about 18,400 jobs. In the IT sector, specifically, that represents a reversal. Previous to these broadband efforts, Kentucky was bleeding IT jobs at a rate of about 6.4 per cent per year. In the last 2 years, we have seen a 4.1 percent increase.

And so that is something that the State is proud of and something that Connected Nation is proud to be a part of as we take this model from State to State, as it is highly transferrable, and we are seeing some early results mirror those Kentucky results in the other States we are in.

When we started in Kentucky 3 years ago, about 60 percent of households had the ability to access broadband. Today, right at about 95 percent of households have the ability to access broadband. Equally important, I would tell you, Mr. Chairman, and you point this out in your blog post with Free Press, that on the demand side—where we need to really pay attention—we have had an 82 percent increase in folks who are actually using the broadband once it is available.

And so as we designed the plan that we have put in place in Kentucky and now in other States like Tennessee and West Virginia, it was with the needs of small business in mind. We looked at the challenges facing small business, and as we all know, so many of the challenges that are faced by entrepreneurs and small businesses are related to isolation. That is so often the reason that they fail. They are either isolated because of their relative size or they are isolated because of their location, isolated from capital or isolated from their potential customers, from market intelligence.

And so we realized that broadband can fix these things, but we also realized that in rural areas, rural States like Kentucky, that problem is two-fold. And so we said we have to help our small businesses. We have to equip, or we have to improve our education providers, our health care providers, and so we developed this plan that was based on a dual approach, a dual focus on both supply and demand.

And so we started out with a map where all providers cooperated and gave us their specific service-level data so that we could understand where those gaps existed, and so then we could drill down into those unserved areas and help providers understand what the market opportunities were in those unserved areas.

At the same time, we worked at the grassroots level. We do work now at the grassroots level with communities and helping build awareness of what are the opportunities related to broadband, why should we be subscribing, and as you point out, Mr. Chairman, that is not a hard sell. These rural communities understand the opportunities associated with broadband.

Bringing those two together, we identify those opportunities for providers. We raise interest, raise awareness, aggregate demand locally. And so we have seen providers recognize those local market opportunities and invest at a rate over the past 3 years in Kentucky that equates to about $700 million in private sector investment. That is an amount that is unprecedented in Kentucky.

And so as we look at the impact, the impact is certainly profound across consumers, across businesses. We see in our business sector when you look at businesses that subscribe to broadband, their revenues are about four times that of businesses that don't subscribe to broadband. Consumers report that they are saving literally billions of dollars a year based on their use of broadband And so to your question earlier, as I am wrapping up here, I would tell you that there are a couple of pieces of legislation that are on the table right now. I would mention Senator Durbin's Connect the Nation Act, which also shares many similarities with Senator Inouye's bill which passed unanimously out of committee, S. 1492, which I appreciate the Chair and the Ranking Member's support on that bill, particularly.

I would say that one of the best things that the Senate could do at this point is to make sure that that bill reaches the desk of the President, and that would enable States to replicate the things that Connected Nation is doing across the country today.

Thank you, Mr. Chairman.

[The prepared statement of Mr. Mefford follows:]

TESTIMONY OF BRIAN R. MEFFORD, PRESIDENT & CEO, CONNECTED NATION, INC.

United States Senate
Committee on Small Business and Entrepreneurship
Honorable John Kerry, Chairman
Honorable Olympia Snowe, Ranking Member

"Improving Internet Access to Help Small Businesses Compete in a Global Economy."

Wednesday, September 26, 2007, at 10:00 a.m.

Chairman Kerry, Ranking Member Snowe, and Members of the Committee:

Thank you for the opportunity to speak with you today regarding the important relationship between broadband Internet and the ability of American entrepreneurs and small businesses to compete in the global economy.

Entrepreneurs and small businesses too often fail in America due to numerous reasons, many of which can be linked to their relative isolation. For those who are working in rural areas, the risks associated with isolation are a double threat:

- Capital is difficult to acquire because they are isolated from fund sources;
- Workers are difficult to find and hire because the demands of the business and the cost of overhead result in relative isolation;
- An entrepreneur's market position relative to established competitors creates something similar to product and service isolation – making it more costly to bring products to market and promote them once they are there;
- Entrepreneurs and small businesses can be isolated from market intelligence and research that otherwise provides a competitive advantage to larger more established companies who are better able to identify customers and target products, services and messages; and
- Finally, with relatively limited resources, American small businesses can be isolated from their own potential customers, unable to spend the dollars necessary to connect with and communicate to those around the world who would otherwise buy their products and services.

Of course, the Internet changes all of this. Broadband Internet practically eliminates the significance of distance, allowing small businesses to break the isolation barriers that have historically placed them at a competitive disadvantage from inception.

With a broadband connection, American business owners can connect literally to the world's resources regardless of physical location – as easily as their larger competitors.

- They can more easily connect with capital resources;
- They can connect to and employ workers regardless of how far apart those workers may be physically;
- Entrepreneurs can bring products and services to market through online resources that provide a global storefront;
- They have equal access to the same quantity and quality of market intelligence; and
- They can identify, connect with, and communicate to customers anywhere in the world.

Connected Nation is a national non-profit that is dedicated to increasing access to and use of broadband in America so that individuals and businesses are better equipped to compete in the global economy. At the state level, we create public-private partnerships that bring together the providers of telecom services and information technology companies with policy makers, local leaders and the consumers of technology to identify the best path to accelerating the availability and use of technology in all local communities.

Connected Nation's proof of concept project, ConnectKentucky, has provided dramatic results that are now being emulated by Connected Nation in other states across the country, including most recently in Tennessee and West Virginia.

- When we began in Kentucky three years ago only 60% of homes could access broadband. Today, 95% have the ability to connect and Kentucky is on track to have 100% broadband availability by the end of this year;
- Home broadband use has grown dramatically by 82%, encouraging private providers to continue their investments in infrastructure statewide;
- Over the last three years, more than 18,400 total technology jobs have been created in Kentucky. In the IT sector alone, Kentucky jobs have grown at a rate four times the national growth rate, representing a reversal from years prior to the ConnectKentucky initiative, when jobs were bleeding out at a decline rate of 6.1%;
- Representing a reversal of the all-too-common rural "brain drain", 96% of Kentuckians who graduate from college remain in Kentucky to live, work and raise their families.
- Kentucky's broadband users estimate they save a total of $1 billion per year; save 230 million hours per year; drive 1 billion fewer miles per year; and report being healthier and better educated as a result of having broadband access;
- Today in Kentucky entrepreneurs are thriving; small businesses are finding an environment ripe for growth; and rural communities are finding ways to diversify and provide attractive opportunities for their children.

These metrics represent a technology turn-around for the Commonwealth of Kentucky. Three short years ago, Kentucky could be found listed at the bottom of nearly every technology-based ranking or new economy index. However, in 2004, we began working aggressively through our public-private partnership to reverse these trends creating a friendly environment for families and businesses eager to excel in the global economy.

We identified that Kentucky's broadband challenge (consistent with the nation's challenge) is not simply an issue related to the *supply* of broadband but one also connected to the *demand* or the use of broadband and related technology. With that understanding, we outlined a course of action that would both enhance the availability of broadband while also dramatically increasing the number of homes and businesses using computers with broadband connections.

First, we established a map-based inventory of all the areas broadband did and did not exist. This inventory was completed through the cooperation of providers who submitted data pertaining to where exactly their broadband services were available. This physical service-level data resulted in an extremely accurate picture of the gaps that existed in broadband service availability.

Once the gaps were identified, we were able to drill down into those unserved areas by gathering additional market intelligence that would bolster the case for providers to extend their services: these data include household density, planned development and other factors such as likelihood that a critical mass would subscribe to broadband service once available.

In conjunction with this "supply side" work, ConnectKentucky also began working locally in each of Kentucky's 120 counties on the "demand side" to create "eCommunity Leadership teams" to identify and generate demand for technology across multiple sectors, including: local government, business, education, healthcare, agriculture, libraries, tourism and non-government organizations. By creating "locally-owned" technology strategic plans, ConnectKentucky was able to deliver additional market-based motivation to would-be providers potentially interested in deploying broadband service in the community.

This dual focus model has worked in rural Kentucky and we're seeing early progress mirrored in other states where Connected Nation has launched similar programs. Today in Kentucky nearly 100% of households have the ability to access broadband which means that, regardless of their physical location, Kentucky entrepreneurs and small business owners are able to connect to the global economy using broadband.

The case is perhaps best illustrated through the story of a Kentucky small business owner named Kamren Colson. As a Kentucky expatriate operating a graphics design firm out of state, Kamren decided to push the envelope related to the promises of broadband technology. Historically, Kamren's family had farmed in the "Burley Belt" of central Kentucky and still owned several acres of land formerly used to raise tobacco. Kamren looked at that piece of land, considered the broadband technology that was available, and decided to relocate his company's headquarters to the serene and rolling Kentucky countryside. With his new business location and a broadband connection, Kamren's business successfully pursued the contract to create and produce promotional pieces for the 2006 Academy Awards in California.

Executives from the Academy reported they never really considered the fact that Kamren and his staff weren't actually "just down the hall" working with them collaboratively. The technology ensured seamless interaction as the two groups collaborated on promotional assets -- trading files and ideas ahead of production. With the brain power of Kamren and his staff, combined with the power of broadband technology, the significance of location was eliminated – the small business was able to work with Academy officials from a former cow pasture in Kentucky just as effectively had it been located down the hall in their Los Angeles office.

For entrepreneurs and small business owners spread across rural America, the challenges are similar and so too are the opportunities. As broadband has become a critical element for success, our nation needs a comprehensive and common sense broadband plan that rewards the innovation of our private sector and creates an environment that is attractive for ongoing investment.

I appreciate the opportunity to speak with you today on behalf of Connected Nation and I look forward to responding to your questions.

Chairman Kerry. Thank you very much for your thoughtful comments. Mr. Levin.

STATEMENT OF DOUGLAS A. LEVIN, PRESIDENT AND CHIEF EXECUTIVE OFFICER, BLACK DUCK SOFTWARE, INC., WALTHAM, MASSACHUSETTS

Mr. LEVIN. Thank you, Chairman Kerry and Ranking Member Snowe. I am the CEO of a Boston area startup company and this particular issue of Internet access for small businesses is particularly poignant because I believe that small companies are impacted by this issue. Larger companies are, in effect, small telecommunication companies. Companies that are publicly-held companies have infrastructure internally, private networks and other means by which to deliver their own infrastructure to their employees, as well as their customers and their partners. And so a small company is impacted by this issue and I am going to give you a couple of examples in my testimony.

I think that the global software industry is changing a great deal and it is impacting the U.S. software industries significantly because one of the shifts in software delivery is software is a service which is highly dependent on the Internet and U.S. companies are operating at somewhat of a disadvantage in offering this new model of software as a service.

Secondly, startups and small- and medium-sized software companies have problems delivering their software and the data and various other parts of their service offerings through conventional Internet connections.

And finally, poor Internet connections in suburban areas and rural areas impact small companies because they can't encourage telecommuters, their employees who are living in rural areas and need to commute in in the eventuality of snow or other issues. Poor Internet connections discourage this telecommuting.

By way of background, I am a 27-year veteran of the software industry. I worked at Microsoft for around 9 years. I have been the CEO of a bunch of Internet startups in the Boston area and I am the CEO of Black Duck Software today. I also served on the Cable Monitoring Committee for the town of Brookline, Massachusetts, where we struggled to introduce two competitors into the marketplace and get Internet access into a community with lots of Ph.D.s, but also lots of people who just demand the Internet access for their families as well as themselves.

Black Duck Software was born out of the idea of realizing that corporations use the Internet as a collaboration medium. Today, we are backed by seven top-tier VC and we are headquartered in Waltham, Massachusetts, and have five offices across the country, as well as offices in Amsterdam and the United Kingdom. We employ 81 people and we have 400 customers worldwide.

The idea for Black Duck was born while I was lying on a beach in Cancun, Mexico, thinking about the problem of exchanging data across the Internet and getting developers to be highly productive. And the reason why I mention that is because inspiration can come in all different ways at different places, and to have universal Internet access is a very important thing in the genesis phase of an entrepreneurial endeavor.

With respect to the changing model of the software industry, software as a service promises to deliver software applications over the Internet inexpensively for small businesses, as well as large businesses and at a fraction of the cost of the conventional applications. It offers a big advantage for small companies and where they can save money, especially on IT infrastructure. Software as a service, however, is Internet intensive, and in the United States today, this is holding back the expansion of software as a service because in some areas of the country, there are people who literally cannot get these applications through their local pipes.

A second issue for Black Duck is we offer lots and lots of updates to our software through the Internet and some of those updates come in the form of software and some of it comes through data. But in either case, we are constantly updating our software, and we need high-speed Internet services to deliver them. Our competitors, who do not have as advanced applications as we do, do it over the Internet. Their applications are smaller. Our applications, because they are so robust, have to be delivered sometimes via the U.S. mail instead of the Internet. This is sometimes hard to comprehend when we are sitting in meetings, but it is a fundamental thing that very advanced technology businesses in the United States are operating at a competitive disadvantage, and you can see it pragmatically day to day in the business when we talk about costs and we talk about delivery and customers.

Chairman KERRY. Is that because of the speed or the volume and size?

Mr. LEVIN. It is both.

Chairman KERRY. Both?

Mr. LEVIN. The pipes are not big enough and the speed is an issue. And by contrast, I could do this in Denmark [snapping of fingers] like that—in the middle of a field. In fact, they have an advertisement where they talk about in rural areas of Denmark you can get 10 gigabytes downloaded to you in the middle of a field.

Poor Internet capabilities in suburban and rural areas make it very difficult for American companies also for this telecommuting issue. It is interesting to note that when I drive by Boston College—I live on Beacon Hill downtown—when I drive by Boston College, which is only a couple of miles away from downtown Boston, my services are not there. They are not available. When I go to the Berkshires for strategic offsites, which are 2 1/2 hours away from Boston, I don't have Internet access. And this is in Boston, and Massachusetts is supposed one of the most advanced States in the country.

Do we work around it? Absolutely, because we are entrepreneurs. However, it makes things more difficult and costly.

So I would urge you to create a national broadband strategy that encourages the creation of a new generation of information superhighway for the new millennium. Thank you very much.

[The prepared statement of Mr. Levin follows:]

Testimony Douglas A. Levin to

United States Senate's Committee on

Small Business & Entrepreneurship

Wednesday, September 26, 2007

Good morning Chairman Kerry, Ranking Member Snow, and other Senators.

I'm pleased to be here to discuss, "Improving Internet Access to Help Small Business Compete in a Global Economy."

There are many issues that are directly and indirectly connected to this subject. But, I would like to focus on three issues related to Internet access for small businesses competing in the global economy:

1. The global software industry is moving toward a bandwidth-intensive Software-as a-Service model.

2. Startups, small and medium-sized software developers, find that it is difficult and expensive to deliver the latest software and data updates via today's conventional Internet connections.

3. Poor Internet capabilities in many suburban and rural areas make it difficult for American companies like ours to support telecommuters.

But before I cover these three issues, let me give you some background about myself and my company.

Black Duck Software - Background

I am a 27-year veteran of the software industry. I started my career working for a government economic development agency in New York City, then I went to work on the Apple Macintosh development team. From 1987-1993 I worked at Microsoft Corporation in various roles. In the mid 1990s I managed a consulting firm, mostly doing projects for Internet startups and telecommunications companies. I was the CEO for two Boston-area Internet startups before founding Black Duck Software, the company I am the CEO of today.

You should also know that I served on the Cable Monitoring Committee for the Town of Brookline, Massachusetts from 1997 to 2001 helping to establish the town's cable Internet access and other services.

Black Duck Software was born when I realized that corporations could use the Internet as a collaboration medium for software development. In other words, I was convinced that companies could achieve greater productivity by building advanced software applications based on software components developed over the Internet in different locations. But these corporations had to track these code components, know their origins, and determine whether they were properly licensed.

I founded a company based on these ideas, and seven top-tier venture capitalists bought into my dream. Today, Black Duck Software is headquartered in Waltham, Massachusetts, and we have offices in five cities across the country, as well as in Amsterdam and the United Kingdom. We employ 81 great people, and we have almost 400 customers worldwide.

How the Internet is driving economic development

The Internet is empowering new businesses, new business models, and global competition. Black Duck Software is at the leading edge of delivering new technology, in new ways, to customers in the technology and enterprise markets.

As I mentioned at the beginning of my testimony, there are three issues that Black Duck's customers face today that are directly impacted by the quality and availability of Internet access:

1. *The global software industry is moving toward a bandwidth-intensive Software-as a-Service model.* Software-as-a-Service promises to deliver software applications over the Internet. This new model enables small businesses to take advantage of sophisticated software applications at a fraction of the cost of conventional applications. Unlike conventional software, this Internet-based software is copied onto a computer's hard drive only when a customer needs it.

   One big advantage of Software-as-a-Service is that small businesses can save money because they don't need their own IT infrastructure. The software takes care of all that. With this new Software-as-a-Service model, small businesses can compete with larger businesses, and more easily and less expensively engage international competition as well.

The United States needs to be leading the way in Software-as-a-Service. But we can only do so if we have enough Internet capacity. If we don't, we will be followers, watching India and China pass us by.

The problem we face in fully executing the Software-as-a-Service model is the stability and speed of the Internet connection. Both these issues affect the user's experience and the software's performance. My company has a Software-as-a-Service offering, and our customers are impacted by these issues.

2. *It is difficult and expensive to deliver our software and data updates via today's conventional Internet connections.*

Black Duck is constantly updating our software solutions for customers. We need high-speed Internet services in order to deliver our updates to customers. Some US customers of Black Duck force us to send updates via the US Postal Service because their Internet service is inadequate for our required download speed. By contrast, our competitors email their software updates over the Internet. As the most advanced technology solution in the marketplace today, Black Duck is sometimes held back by bad Internet services."

Furthermore, today's advantage in software development is gained through taking pieces of software code located in various software repositories across the Internet and combining them into one software application. Everyone in the software development community today is doing this. It lowers the cost of development and enables information and data to be shared across partners, vendors and suppliers. This drives responsiveness to customers. But we need your help to drive down the cost and availability of the next generation of Internet infrastructure.

3. Poor *Internet capabilities in many suburban and rural areas makes it difficult for American companies like ours to support telecommuters.*

By contrast, it is easier for people to telecommute in the Netherlands, Denmark and other countries.

It is difficult to locate businesses or workers outside of locations close to major metropolitan areas, such as western Massachusetts. Lack of bandwidth makes collaboration much more challenging.

Even in metropolitan areas, for example, the quality of human conversations (so-called Voice-over-IP) is quite low when made over an Internet connection. There are many gaps in the conversation due to data loss.

In Denmark: Download speeds can be 30x faster and uploads 200x faster than the US for rural Internet access.

The Internet is fueling innovation around the world. It was developed here in the United States, and American companies like Google have become household names from Milan to Moscow and Bonn to Beijing.

But other nations are catching up fast. They understand that their future economic prosperity depends on harnessing the power of the Internet to create new jobs for their people. So while they invest billions in infrastructure, America is in danger of falling behind. My point is that the health of the American economy depends in no small measure on the health of our Internet infrastructure. The jobs of tomorrow depend on the technology decisions this august body makes today.

President John F. Kennedy once challenged Americans to "take longer strides" and lead the world by putting a man on the moon – and we did it. Today I'd like to challenge you to take some longer strides of your own. I urge you to create a national broadband strategy that encourages the creation of a new generation of the information superhighway for a new millennium.

We have the talent to lead the world in the 21st century. We have the ideas and a vision of a better world for our children and grandchildren that technology will help bring into being. We have the entrepreneurial wherewithal to continue to lead the world with new products, services and businesses. We need the conduit to get us there.

Thank you again for the honor of addressing you today.

Chairman KERRY. Very helpful. Congratulations on what you are doing. That is a very interesting perspective for us to hear.

Dr. Wallsten.

STATEMENT OF DR. SCOTT WALLSTEN, SENIOR FELLOW AND DIRECTOR OF COMMUNICATIONS POLICY STUDIES, THE PROGRESS AND FREEDOM FOUNDATION, WASHINGTON, DC

Dr. WALLSTEN. Mr. Chairman and Senator Snowe, thank you for inviting me here and giving me the opportunity to testify. I will make three points.

First, there is not an overall U.S. broadband problem. Telephone, cable, and wireless companies are investing billions in new highspeed infrastructure. Consumers and businesses are adopting broadband at remarkable rates.

Second, those who believe there is a problem advance proposals that sound appealing, but they don't demonstrate that their proposals would actually benefit consumers and businesses.

Third, despite substantial current investment, policies can still affect broadband's growth. In particular, we need to collect better data that would allow us to rigorously analyze proposed policies and to remove arbitrary barriers to entry that continue to prevent the market from reaching its full competitive potential. Government could help achieve both goals. I will elaborate on those points.

First, the sky isn't falling. There is scant evidence of a U.S. broadband problem. Nearly half of all American households subscribe to high-speed Internet connections, more than twice as many as just a few years ago, and about 60 percent of businesses with fewer than 100 employees have broadband connections.

Earlier this month, the National Federation of Independent Businesses reported the results of a survey that asked members to state their most important problem. Broadband didn't make that list.

Internet service providers are investing in broadband infrastructure at unprecedented rates. Cable countries are expected to invest about $15 billion this year upgrading their networks. Verizon alone is planning to spend $23 billion on its fiber optic network by 2010. By the second quarter of 2007, its fiber services were available to nearly 8 million homes and are expected to reach 9 million by the end of the year. Cellular mobile companies continue to upgrade and build high-speed networks while other firms are building out new wireless networks that offer coverage ranging from very local to national.

But supply is not the only factor that affects the state of broadband. Demand is also crucial in determining broadband penetration and speeds. I understand that some advocates think faster is always better. Like them, I live online and place a high value on a very fast connection. But not everyone has the same preferences that we do. Few small businesses, for example, download multiple movies every day or engage in bandwidth-intensive online gaming. Many people in small businesses are simply unwilling to pay more for higher speeds. That is why not everybody signs up for the fastest speed they can get.

Those who believe the United States has a broadband problem claim that broadband speeds in the United States are much slower than elsewhere. These claims are simply wrong. They are based on comparisons of advertised, not actual, speeds. According to speedtest.net, which has data from nearly 200 million unique speed tests of actual broadband connections around the world, the average U.S. speed ranks about third or fourth globally.

In short, the evidence contradicts the argument that there is too little investment in broadband infrastructure or that most consumers and small businesses are desperate for more. The important question is whether market failures or other obstacles hinder broadband investment, competition, and adoption by consumers and businesses. Because investment dollars are scarce and because policies have costs as well as benefits, we should analyze policies carefully and rigorously to ensure that their expected benefits exceed their expected costs. Unfortunately, few proposals are accompanied by analysis.

For example, many who believe the United States has a broadband problem argue that France and Japan are doing well because they require their biggest telecom companies to open their infrastructure to competing broadband providers. This regulation is known as unbundling, which is sort of like making Starbucks lease space and equipment to any free-lance barista who stops by. But the truth is more subtle. France does not apply unbundling regulations to fiber optic lines, and in Japan, the regulated price for a firm to use the fiber is so high that essentially no company takes advantage of that regulation. Instead, the incumbent telephone company and the electric power utilities are building and operating fiber themselves. In other words, unbundling proponents point to Japan and France as models to emulate, but those countries have, for all practical purposes, not applied unbundling to the very type of infrastructure those proponents want to see here.

As another example, some argue that expanding the Universal Service Fund to include broadband services might benefit small businesses. But expanding that fund is more likely to harm small businesses since they, like all other consumers, pay for universal expenditures through taxes on their own telecom services. That is why the National Federation of Independent Businesses argues strongly against increasing the fund.

I do not, however, intend to imply that the market is perfect. We know that the overall positive picture of broadband in the United States can mask underserved geographic areas and socioeconomic groups. Data collection efforts should be targeted at identifying potential problems and at gathering the information necessary to evaluate whether proposed policies are likely to address them effectively. That is why models like Connect Kentucky are successful. They carefully identify areas where there might be a problem and help tailor specific solutions.

In addition, certain regulations continue to make it more expensive than necessary for new companies to enter the market. For example, there is no economic justification for requiring a special license or franchise to offer cable television services over broadband lines.

And despite strong investment in wireless networks, hundreds of megahertz of spectrum remain unused or inefficiently used by the private sector and by the Government. Every day that spectrum re mains unavailable for high-value use represents a tremendous opportunity cost, a significant loss to our economy.

To conclude, let me reiterate that the key to good broadband policy is careful analysis that attempts to identify market failures or artificial barriers suppressing broadband investment and adoption, followed by rigorous evaluation of whether proposed interventions are likely to yield net benefits. And precisely because the Internet is so important, Congress should be cautious and consider carefully interventions in this fast-changing industry to ensure that they do not unintentionally reduce incentives to invest in the very infrastructure we all believe is so important. Thank you.

[The prepared statement of Dr. Wallsten follows:]

Statement of Scott Wallsten, Ph.D.
Senior Fellow and Director of Communications Policy Studies
The Progress & Freedom Foundation

Communications, Broadband and Competitiveness

Before the
Committee on Small Business and Entrepreneurship
U.S. Senate

September 26, 2007

Mr. Chairman and members of the Committee, thank you for inviting me here and giving me the opportunity to testify. I will make three points.

First, there is little evidence of a U.S. broadband problem. Telephone, cable, and wireless companies are investing billions in new high-speed infrastructure, and consumers and businesses are adopting broadband at remarkable rates.

Second, those who believe there is a problem advance proposals that sound appealing, but they fail to provide solid analysis showing that their proposals would actually benefit consumers or small businesses.

Third, despite significant infrastructure investment, we can do better. In particular, we need to collect better data that would allow us to rigorously analyze proposed policies and to remove arbitrary barriers to entry that continue to prevent the market from reaching its full competitive potential. Government can help achieve both goals.

I'll elaborate on those points.

First, the sky isn't falling. There is scant evidence of a U.S. broadband problem. Nearly half of all American households subscribe to high-speed Internet connections, more than twice as many as just a few years ago.[1] About 60 percent of businesses with fewer than 100 employees have broadband connections.[2] Earlier this month the National Federation of Independent Businesses reported the results of a survey that asked members to state their most important problem.[3] Broadband did not make the list.

Internet service providers are investing in broadband infrastructure at unprecedented rates. Cable companies are expected to spend about $15 billion this year upgrading their networks.[4] Verizon alone is planning to spend $23 billion on its fiber-optic network by 2010.[5] By the second quarter of 2007 its fiber services were available to nearly 8 million homes, and are expected to reach 9 million by the end of the year.[6] Cellular mobile companies continue to upgrade and build high-speed networks, while other firms are building out new wireless networks that offer coverage ranging from very local to national.[7]

Supply is not the only factor that affects the state of broadband. Demand is also crucial in determining broadband penetration and speeds. I understand that some advocates believe faster is always better. Like them, I live online and place a high value on a very fast connection. But not everyone has the same preferences that we do. Few small businesses, for example, download multiple

---

[1] http://www.pewinternet.org/pdfs/Broadband_Commentary.pdf
[2] IDC market analysis, March 2007. "U.S. Small Business Internet 2007-2011 Forecast."
[3] http://www.nfib.com/object/IO_34726.html
[4] http://www.infonetics.com/resources/purple.shtml?msna07.cpx.2h06.nr.shtml
[5] http://policyblog.verizon.com/policyblog/blogs/policyblog/czblogger1/290/fios-fact-sheet.aspx
[6] http://investor.verizon.com/financial/quarterly/vz/2Q2007/2Q07Bulletin.pdf
[7] See, for example, http://www.pcmag.com/article2/0,1895,2186108,00.asp or http://www.believewireless.com/.

movies every day or engage in bandwidth-intensive online gaming. Many people
and small businesses are simply unwilling to pay more for higher speeds. That's
why not everybody signs up for the fastest speed they can get.

Those who believe the U.S. has a broadband problem claim that
broadband speeds in the U.S. are much slower than elsewhere. These claims
are simply wrong. They are based on comparisons of advertised, not actual,
speeds. According to speedtest.net, which has data from nearly 200 million
unique speed tests of actual broadband connections around the world, the
average U.S. speed ranks about third or fourth globally (Figure 1).

Figure 1
Average Actual Broadband Connection Speeds Across Countries

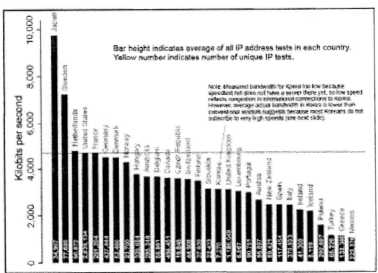

Source: Speedtest.net. Average of tests from August 2006 – June 2007.

In short, the evidence contradicts the argument that there is too little investment in broadband infrastructure or that most consumers or small businesses are desperate for more.

The important question is whether market failures or other obstacles hinder broadband investment, competition, and adoption by consumers and businesses. Because investment dollars are scarce and because policies have costs as well as benefits, we should analyze policies carefully and rigorously to ensure that their expected benefits exceed their expected costs. Unfortunately, few proposals are accompanied by serious analysis. For example, many who believe the U.S. has a broadband problem argue that France and Japan are doing well because they require their biggest telecom companies to open their infrastructure to competing broadband providers. This regulation is known as unbundling, which is sort of like making Starbucks lease space and equipment to any freelance barista.

The truth is more subtle.

France does not apply unbundling regulations to fiber optic lines. And in Japan, the regulated price for a firm to use the fiber is so high that essentially no company takes advantage of the regulation. Instead, the incumbent telephone company and the electric power utilities are building and operating fiber networks themselves. In other words, unbundling proponents point to Japan and France as models to emulate, but those countries have, for all practical purposes, not applied unbundling to the very type of infrastructure those proponents want to see here.

As another example, some might argue that expanding the Universal Service Fund to include broadband services might benefit small businesses. But expanding the fund is more likely to harm small businesses since they, like all other consumers, pay for universal service expenditures through taxes on their own telecommunications services. That's why the National Federation of Independent Businesses argues strongly against increasing the fund.[8]

I do not, however, intend to imply that the market is perfect. We know that the overall positive picture of broadband in the U.S. can mask underserved geographic areas and socioeconomic groups. Data collection efforts should be targeted at identifying potential problems and at gathering the information necessary to evaluate whether proposed policies are likely to address them effectively. That's why models like ConnectKY appear to be successful—they carefully identify areas where there might be a problem and help tailor specific solutions.

In addition, certain regulations continue to make it more expensive than necessary for new companies to enter the market. For example, there's no economic justification for requiring a special license to offer cable television services over broadband lines.

And despite strong investment in wireless networks, hundreds of megahertz of spectrum remain unused or are used inefficiently by the private sector and by the government. Every day that spectrum remains unavailable for high-value uses represents a tremendous opportunity cost—a significant loss to our economy.

---

[8] http://www.nfib.com/page/technology.html

To conclude, let me reiterate that the key to good broadband policy is careful analysis that attempts to identify market failures or artificial barriers suppressing broadband investment and adoption, followed by rigorous evaluation of whether proposed interventions are likely to yield net benefits.

And precisely because the Internet is so important, Congress should be cautious and consider carefully interventions in this fast-changing industry to ensure that they do not unintentionally reduce incentives to invest in the very infrastructure we all believe is so important.

Thank you.

Chairman KERRY. Thank you.

Well, we seem to have not just a disconnect out in the country at large, but we also have a disconnect between you and Mr. Levin right here, so let me feel this out a bit. Are you satisfied with the United States going backwards in terms of other countries?

Dr. WALLSTEN. Well, I think that the rankings are actually not very useful at all and there are many reasons not to pay attention to simply just rankings and not use them as a basis to make policy.

First of all, the data that the OECD puts out themselves are very problematic. They——

Chairman KERRY. You use that data in your own charts.

Dr. WALLSTEN. The data in the chart in this figure is from speedtest.net. But——

Chairman KERRY. No. In addition to that, don't you have some other—I thought you had some additional data there.

Dr. WALLSTEN. I don't believe I used data from the OECD in this paper, but I actually have used the data from the OECD in papers and the way that I use the data and the way that I think the data should be used is to control carefully—control for things that policy can't affect, like population density. That is not offered as an excuse, it is simply an empirical fact. Every single empirical study on broadband penetration finds that population density is correlated with it. Control for things like that and test for the effects of factors that policies can affect. Then you are not looking simply at rankings, you are controlling for lots of things.

I mean, it doesn't make sense, for example, to compare the United States to Iceland, which ranks third in the OECD rankings, since Iceland has a population of 300,000, which might compare to Buffalo.

Chairman KERRY. Dr. Wallsten, it is a relative deal if some countries are bigger than other countries. But if the country's population as a whole has access and they are all able to use it, that is one measurement, isn't it?

Dr. WALLSTEN. Well, that is right, and that is why I think it is important also to look very carefully——

Chairman KERRY. Dr. Wallsten, this is your chart here, and broadband subscriptions per capita by technology, it says, Scott Wallsten——

Dr. WALLSTEN. That is right, and what else is on there? Chairman KERRY. OECD.

[The chart being referenced follows:]

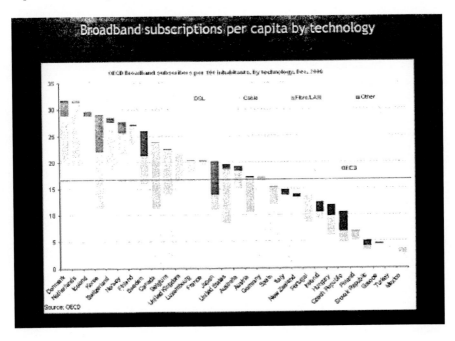

Dr. WALLSTEN. Umm-hmm, and what is the heading on the chart?

Chairman KERRY. Broadband subscriptions per capita by technology. So you are using, I guess, the OECD——

Dr. WALLSTEN. That is true, but the point in that one is to show the mix—

Chairman KERRY. So it is selective. You use it where you want to and——

Dr. WALLSTEN. No. Senator Kerry, I am sorry, that is not correct. I try to use that data appropriately, and the data themselves—I am not trying to make excuses for the United States. I interested in using the data appropriately. The data——

Chairman KERRY. Just help me understand it. Mr. Levin, who is in business, has described a situation where he can't achieve his business goal because we don't have adequate capacity. But he can achieve it in another country. Isn't that an incentive to go and operate out of the other country?

Dr. WALLSTEN. Well, I would prefer not actually to use anecdotes as basis for making policy.

Chairman KERRY. But that is real life.

Dr. WALLSTEN. No, Senator Kerry, the OECD data omits, for example, all university connections. It omits most——

Chairman KERRY. I am not talking about OECD now. I am talking about the practical reality of speed and access——

Dr. WALLSTEN. The question for any policy is whether its expected benefits exceed its expected costs, and it is possible you could pass a law that would mandate, for example, a minimum speed for broadband that would be very high and that might aid his company. The cost for that might be very high, and that is a question you want to ask. What are the costs of a proposed policy expected to be? Right now, we don't even have the data to be able to answer that question well.

And I do have—I mean, there are other suggestions of things that we can do. I think there are things we can do right now to improve the broadband situation—

Chairman KERRY. What are you suggesting? There is something I don't understand here. I mean, a community ought to have access to broadband and be able to make the choice within the community of whether you want to buy, at what speed you want to buy, et cetera.

Dr. WALLSTEN. Exactly. People should be able to choose the speed they want to buy.

Chairman KERRY. But you have to have that availability to be able to do it and right now we don't have that availability.

Dr. WALLSTEN. But that doesn't mean that everybody should invest, every community should automatically invest in 100 megabit per second availability. They have other priorities, I am sure.

Chairman KERRY. Mr. Levin, what do you say to that?

Mr. LEVIN. I disagree with a lot of the points that he has made during the course of his testimony. I think even in the most concentrated areas of technology, like for example, the Silicon Valley, and also Massachusetts, it is difficult sometimes to find wireless connections, good Internet connections, and building a business has some fundamental challenges connected to it and getting inexpensive broadband to a small business is challenging in the United States today.

Dr. WALLSTEN. If I could just—could I just follow up for 1 second? There are things that Congress could do right now. The AWS spectrum auction concluded more than a year ago. Companies spent billions of dollars on spectrum. For example, T–Mobile, Leap Wireless, Metro PCS, Comcast all bought spectrum hoping to build out broadband networks. Many of them are having trouble because the government agencies that were on that spectrum are not moving away.

That is something that Congress could do right now to open up more wireless space for broadband. That doesn't require a summit, a broadband summit. There are wireless opportunities that we could be doing right now, and those would be great for improving competition.

Chairman KERRY. So you disagree with the President's goal that we ought to have ubiquitous broadband——

Dr. WALLSTEN. I think we ought to make sure that we do everything we can to make sure that the market is competitive.

Chairman KERRY. Do you think we have done everything we can to make it competitive?

Dr. WALLSTEN. I think there are things that we should be doing. I think franchise regulations are serious impediments to firms investing. One thing that we don't pay very much attention to is demand. One of the reasons that consumers in France and Japan, for example, would buy higher-speed connections is because companies have always been allowed to offer video—television video—over broadband lines. Here, you can't do that without a franchise and there is not—I mean, I understand there are fiscal reasons why cities need those franchise rules, but there is not an economic reason for that and without being able to purchase cable television services over broadband, that reduces demand.

Chairman KERRY. Mr. Scott, what is your reaction to this?

Mr. SCOTT. Well, I have no doubt that Dr. Wallsten comes by his opinions honestly and some of his critiques in his academic papers I find interesting. I disagree with most of them, but I think his analysis is worthy.

I look at the debate over the broadband problem over the last few years and it reminds me somewhat of the global warming debate. The overwhelming amount of evidence is on one side, as far as I can see, and the telephone companies, like the oil companies, can make a really nifty PowerPoint presentation to provide the opposite, but it doesn't make it so. And if we have got evidence from the OECD, the ITU, and Point Topic, and the FCC and numerous other data sources, as well as every foreign telecommunications service provider that is, I think, not lying about the advertised rates of service, I just have to say the broadband problem is very real. It is both about a lack of availability and a lack of competition. That means lower speeds and higher prices. And if we don't do something about it, we are going to suffer economically over the next 10 to 20 years.

Chairman KERRY. Speaking of global climate change, I am Chairing the Foreign Relations Committee meeting with foreign ministers on that subject in about 5 minutes, so I have got to run and do that. But let me just say from our own experience—Dr. Wallsten, you need to sort of know this and then maybe you can respond afterwards for the record—in the Berkshires in Massachusetts, we have a very thoughtful, well-educated economic base which has been handicapped by virtue of the lack of access to broadband. We had to create something called Berkshire Connect to create a consortium to pull various people together in order to create the economic clout to even get people to bid, because they wouldn't bid. They just didn't think there were enough folks there. There wasn't enough money to be made. They wanted to hook up all the big buildings in downtown Boston and other communities first. So there is a race to the easy money, not necessarily a race to where it has social impact.

So this question of utility, of public utility and which comes first, the chicken or the egg here, is a critical one from a public policy point of view. Those schools need access. Kids need access. People need access. We need to educate people about why access is, in fact, good. If you just leave it out there and nobody is aware of what the benefits may be, they may not demand it. But as they become more aware of the benefits and the economic upside in some of the ways that Mr. Levin and others have described, there are all kinds of benefits.

It is hard to ignore a study that says we are leaving 1.2 million jobs and $500 billion off the table because we are not getting that kind of access to high-speed Internet.

Dr. WALLSTEN. And that is why I believe that models like Connect Kentucky are good, because they identify very specific problems. Also, those studies that you cite, the $500 billion one from about 4 years ago, I believe, and the more recent one from Brookings, don't advocate any of the policies that some here have recommended. And I am—all my work is empirical, data-driven, and that is why I think the data is important.

Chairman KERRY. Listen, I am not trying to fight with you, I am just disagreeing with some of your conclusions. But I think it is important to have the testimony. It is important to have the discussion. We wouldn't have invited you here if we didn't think it was important. I think there is a very powerful argument for why, in fact, this access and the competition is so critical.

I am sure that Senator Snowe will further examine that, so why don't I turn it over to her and you can close it out. Thank you.

Senator SNOWE. [presiding.] Thank you, Mr. Chairman.

Dr. Wallsten, so you don't think that there should be any national policy with respect to the broadband deployment, is that right?

Dr. WALLSTEN. Well, I think we need to be careful about what exactly that means. I mean, our data collection right now is very poor. I think everybody has agreed with that. And that is certainly a good place to start.

Senator SNOWE. So if the FCC changes its methodology and the type of data it acquires, which needs to be done soon, and it reaffirms the dramatic problem that we are facing in this country, would you feel differently?

Dr. WALLSTEN. Absolutely.

Senator SNOWE. You would?

Dr. WALLSTEN. I would like to see studies—I would like to see proposed policies and analyses of the expected costs and benefits of those proposals and then we would go from there. I mean, what is sort of amazing to me is that in almost every other area of policy— you think of labor policy, for example, or environmental policy, for example—those agencies collect tremendous amounts of data and policies are based on extensive, careful analyses. And here, for broadband policy, an industry that affects so much of our economy, we want to make policy based on simple rankings that don't provide any sources, that don't tell you what their methodology is, that leave out huge categories of connections. To me, that is simply irresponsible.

Senator SNOWE. And the FCC has acknowledged that their methodology is wrong, correct?

Dr. WALLSTEN. Oh, yes, and——

Senator SNOWE. And that was affirmed by the GAO——

Dr. WALLSTEN. Right. And even the FCC staff know this, too, and would like to work on that problem.

Senator SNOWE. Right. Exactly. Better data is obviously critical, to get our arms around the data and study exactly what the picture of America looks like. But I think the real question is whether or not you can give impetus to the deployment of broadband and what role the Federal Government plays.

I am impressed with Mr. Mefford and what is happening with Connected Nation and Connect Kentucky. Maine has a Connect Maine initiative and I hope it will share the same success. But they have undertaken it because there is a huge vacuum in leadership, even at the national level. These are programs undertaken by local governments that otherwise could not afford to do them, but they recognize it is an economic imperative, especially in rural America.

I mean, that is the real issue here, how we are going to rebuild rural America at a time in which we are dramatically losing manufacturing jobs. In our State, we have lost 17 percent of the manufacturing jobs since 2000. It keeps happening. It happened again recently. We keep losing major companies in rural America. How do you rebuild it? You rebuild it by giving them access to the technology so that they can conduct their small enterprises in these rural economies. You shouldn't have to be in urban America. I think that is one of the real issues that we have to confront in this country today is what we are going to do to assist small towns to rebuild their economies and this is one dimension of that.

I don't know—Mr. Mefford, maybe you can add to this debate about what pace you would expect to happen in other places as compared to Connect Kentucky. The President set a goal in 2004 that by the end of 2007, we would have broadband deployment. That hasn't happened. So what would it take to apply your model across this Nation? How long would it take?

Mr. MEFFORD. Well, first of all, it requires something like S. 1492 or Senator Durbin's Connect the Nation Act. That is what enables States, that will empower States to replicate this model.

Senator SNOWE. You need the broadband mapping.

Mr. MEFFORD. Well, that is the starting point. Somebody said the mapping is sort of like putting on your clothes to go to work. I mean, that is what gets us started. That is what starts this market-based approach that embraces all providers. When I say market-based, I mean that it is this dual focus on both supply and demand, but it is inclusive of all types of providers. And so in Kentucky, when I say we have gone from 68 percent to 95 percent, that includes cable and DSL and fixed wireless and municipal wireless and municipal cable, all these different types of services.

So to answer your question, once that empowering piece of legislation is passed, then the process can begin immediately. We are engaged with about a dozen different States on different levels, so the interest is there and we certainly have the capacity to engage additional States. But once the funding is in place, that certainly, like most things, is the largest impediment.

Senator SNOWE. Do you see that as an appropriate role for the Federal Government?

Dr. WALLSTEN. I think what Connect Kentucky—I think that general approach seems to be exactly right. I mean, they carefully identify where there are problems and then figure out ways to solve them.

Senator SNOWE. But the broadband mapping legislation, for example—Dr. Wallsten. I think that is worth considering.

Senator SNOWE. Would that be enough once that was concluded, how long would it take to have that ripple effect across America?

Mr. MEFFORD. We are talking a matter of months. I mean, if we can establish a single clearinghouse where that data is placed, then it is a matter of processing data and distributing that throughout the country.

Senator SNOWE. Mr. Scott, what is your reaction to that?

Mr. SCOTT. I agree with all these guys that data is an important step. The better data we have, the better policies we can make. I think the partnership between Federal data collection, where you have got a baseline standard that makes tools for organizations like Connect Kentucky to use at the State and local level is the right approach. I think S. 1492 is a good bill. We supported it from its inception.

But I think having the data begs the question that we have issues we need to look at, and in my written statement I have laid out a number of pieces of policy which we think will go toward solving the problem, some small, some large, some that the Commission will do, some that the Congress should do, and I think that we have an opportunity now in the next 12 months to really think carefully about what steps we want to take and what goals we want to reach, because ultimately all policy is made to reach some big picture goal, and if our big picture goal is just to incrementally improve our broadband market, that is one set of policies. If our big picture goal is to produce a world-class infrastructure and duplicate the same kinds of successes we had with electrification and the highway programs, well, that is a different set of policies. I think there is honest disagreement about what you want to do, but you have got to make those choices.

Senator SNOWE. This is why we wrote the Telecommunications Act of 1934. We thought it was in the national interest to extend telephone service to all parts of America. That is why Senator Rockefeller and I created the E-rate program.

Mr. SCOTT. That reminds me of a statement that Congressman Ed Markey, the Chairman of the Telecom Subcommittee in the House of Representatives, said to me once. He said the 1996 Telecommunications Act was a great idea. I sure wish somebody would try to implement it.

[Laughter.]

Senator SNOWE. Good point. And look at where we are today vis- a-vis that policy and how much has dramatically changed. I think it just tells you what the landscape looks like and that is why small enterprises and rural America are struggling with the current market plan. We didn't even factor in wireless at that point. Even with respect to the E-rate program, it was just on the cusp of being discussed and wireless wasn't really part of the picture at that point when we rewrote the Act in 1996.

Mr. Mefford.

Mr. MEFFORD. Senator Snowe, I would say that that point just provides more additional merit for this approach to empower States. I think where States have been active in engaging providers in the context of telecommunications reform, I think we have seen some positive results. Certainly and obviously that hasn't been complete and total or we wouldn't be here today, but again, as we have employed this market-based approach in the States that we are engaged with, and I will reference Kentucky specifically, we have seen that increase and that has been primarily by private sector providers, not totally, but that investment has been made in large part by private sector providers.

In the remaining 5 percent that we have to cover—Kentucky will be at 100 percent broadband coverage by the end of this year. That has required a more entrepreneurial approach and so that does get us to the point where we have to look at things like public-private partnerships that incent investment. And so we may have local governments partnering with private sector providers to build out infrastructure and sharing revenue. But that, again, has been the minority part of our approach.

Senator SNOWE. But there would be a public commitment. Is there a public commitment currently on the Connect Kentucky.

Mr. MEFFORD. Yes, ma'am.

Senator SNOWE. Do the State and local government participate financially?

Mr. MEFFORD. They do. The largest commitment comes from the State, and so generally in our States, about 80 percent of the commitment comes from State government and the remaining 20 percent comes from the private sector, and not just telecommunications providers, but companies in general that have a vested interest in the growth of technology. So we have health care companies and automobile companies and banks and so forth.

Dr. WALLSTEN. If I could just jump in for 1 second, I just wanted to add that I think one of the great things that has come out of their initiatives is a tremendous amount of data that actually will begin to allow us to test the effects of different policies. I know I am very much looking forward to using it.

I would also just like to sort of add on a personal note that that presentation of mine that Senator Kerry was referring to is available on the Web site of the Progress and Freedom Foundation, and I hope that people will look at it to realize how I am analyzing the data and not using data where it seems it helps my case and not. Senator SNOWE. We won't. We appreciate that.

Dr. WALLSTEN. I hope people look at it.

Senator SNOWE. Do you think that there is competition in the market now? Do you think that that is the essence of the problem, as well?

Dr. WALLSTEN. I think it depends where you look. I believe, overall, there is competition. So, for example, there was an article in the Wall Street Journal two or 3 days ago noting that broadband satellite services are becoming faster and cheaper and that is available everywhere in the United States. Verizon and Sprint both offer wireless broadband services. That is generally slower than wired services, but is getting faster.

Senator SNOWE. But isn't it a question of cost?

Dr. WALLSTEN. None of those are reflected on that map. I am sorry.

Senator SNOWE. Isn't it the question of cost?

Dr. WALLSTEN. And those costs are coming down. But you are right. It is a question of cost, and also as we move more spectrum into the market and as there are more options, I would expect those prices to continue to come down.

Senator SNOWE. And the special access decision that will be made by the FCC, do you think that would help to promote growth in competition?

Dr. WALLSTEN. Special access is another complicated question where also the GAO—actually, we would probably be having a very similar discussion if it were just on special access because the GAO's main conclusion was that there wasn't enough data to do an analysis. I would hope that all the various players would come to the table and show their data, because none of the CLECs make their data available, for understandable reasons, and the incumbents don't want to make more available than they are required to and it is very hard to make decision under those circumstances.

Senator SNOWE. And unbundling, do you think that it has helped to open markets, because there has really been a lot of problems with incumbents pulling out of the residential broadband market.

Dr. WALLSTEN. Right. Well, that is slightly a little bit different from unbundling policies. I am actually working on a paper right now, or revising a paper right now, on bundling policies across OECD countries and it didn't work here. Like I mentioned, in France and Japan, unbundling doesn't apply to the fiber lines and so companies are investing in their own fiber optic lines.

One question I have, for example, in Japan, one of the main providers of high-speed service is the electric utilities, not through broadband over power lines, which seems to be next year's technology and always will be, but actual fiber optic connections. Why aren't companies like that doing it here? Why don't electric utilities do it here? Maybe it is a bad business decision. Maybe regulations don't easily allow them to enter other electricity markets—I am sorry, markets other than electricity. I think things like that are worth looking at. I don't know the answer.

Senator SNOWE. Mr. Scott

Mr. SCOTT. I think both the points you raised are critically important. I will start with the current proceedings at the Commission. It is not just special access. It is also forbearance petitions and copper retirement. These are technocratic issues that are very complicated in the regulatory proceedings, but their outcomes will be hugely important in determining the prices and choices that small businesses have, particularly as they grow, and I think we would do well to pay close attention to what the Commission is going to do on those issues.

As far as unbundling goes, I shared Congressman Markey's comment in jest, but I think his point is very valid here. You know, unbundling was never properly implemented in the United States for a variety of reasons, which we can debate at length at another time. But I think if you look at the way unbundling policies have been executed, if you just as a tourist stroll around in any European capital, you will find half-a-dozen or more storefront shops trying to sell you DSL. It is a competitive market the likes of which is impossible to imagine in the United States, and I think that disparity is something that we have got to address. I am not saying that I have the answer chapter and verse today, but I think taking unbundling and putting it back on the table for serious consideration is a very wise move.

Senator SNOWE. I appreciate that. Any other comments? Mr. Levin, do you think we should have a national policy?

Mr. LEVIN. Absolutely. The national policy may address these issues of have and have nots and I would encourage my colleague over here to go out and see some of these disparities that you see in different markets. I travel extensively. I go to South Korea and Japan and I have been all over Europe and I feel like I may not have second-class services, but I certainly don't have first-class services. That affects my business. It affects lots of other businesses. And I can think of my kids in school or people in hospitals. Those are two areas, schools and hospitals, where I think Internet service would actually lower the cost of operating those entities and also offer much more data to people and applications that need data.

Senator SNOWE. I appreciate it. Mr. Mefford, you have the last word.

Mr. MEFFORD. Senator Snowe, thank you again for the invitation today. I would just end by reiterating the fact that America's broadband challenge is as much about demand as it is supply in my mind, and I think Senator Kerry has affirmed this today, affirmed it yesterday in his blog post, to note that we have to acknowledge that the number of people actually using the technology that has already been deployed is extremely low from the household standpoint, and what our model and the data that we have generated after the fact has revealed is that as we can increase those numbers of people actually subscribing to broadband, then providers are obviously more interested in deploying further and further and further out into those developing markets.

The things that can be done at the very grassroots level are basic in nature, but it is about generating awareness and helping households and small businesses understand and appreciate better the value of broadband. We have recently gathered some data where we asked, what are the factors that caused you to become a new subscriber to broadband? The top reasons are things like, well I realized that broadband was worth the extra money. Then there is, I learned that broadband became available in my area. So you can see these are awareness building issues. And the third is, I got a computer in my home, and so we know that is an obstacle that we have to address. On down the list is the point that, well, I decided that broadband became affordable.

So contrary to some conventional wisdom that is out there, our biggest challenge to adoption is not price. It is in raising awareness and improving the value proposition that allows individuals, families, and businesses to make the decision to spend the money on broadband.

Senator SNOWE. But you wouldn't disagree that price is a barrier in many cases?

Mr. MEFFORD. Price is absolutely a barrier for some segments of the population——

Senator SNOWE. Such as those that only depend on one provider right?

Mr. MEFFORD. That is right, but Senator, what we have seen in Kentucky is that what we do, in effect, is lower the cost of entry for new providers or for existing providers to extend their networks. As that has happened, we have seen that now the majority of Kentuckians have a choice between at least two providers. Many have a choice between three and four and five providers, and as that has happened, we have seen the effects of competition and prices have come down.

Far and away, price is not the top reason given that people aren't investing in broadband, but absolutely, to your point, we are focused on addressing price, and we don't consider a broadband solution an option worthy of mapping until it is affordable.

To the point of computers being an obstacle, we have developed programs, again, that are State-specific, one that we called "No Child Left Offline" that actually uses donations from companies like Microsoft and Lexmark and CA and Intel and we put computers in the homes of identified families, underprivileged families, and that addresses that barrier of computer ownership and allows them then to become a broadband subscriber.

Senator SNOWE. Well, I appreciate it. I thank you all very much. It has been very helpful and very critical to this issue, and I thank you for your excellent testimony.

Before we adjourn, we will leave the hearing record open for 2 weeks for additional questions and testimony.

With that, the hearing is adjourned.

[Whereupon, at 11:48 a.m., the Committee was adjourned.]

APPENDIX MATERIAL SUBMITTED

RESPONSE TO POST-HEARING QUESTIONS FROM SENATOR SNOWE

Questions for the Record from Senator Olympia J. Snowe, Ranking Member
Senate Committee on Small Business & Entrepreneurship
to FCC Commissioner Michael J. Copps
Hearing on "Improving Internet Access to Help Small
Business Compete in the Global Economy"
September 26, 2007

1.   In Mr. Wallsten's testimony, he states that expanding the Universal Service
Fund to include broadband services is likely to harm small businesses because they
pay for universal service expenditures through taxes on their own
telecommunications services. In fact, he cites an argument by the National
Federation of Independent Businesses, which opposes increasing the fund because of
its affect on small businesses. During your testimony you note that the Joint Board
on Universal Service is talking about covering broadband with universal service. If
these changes are made, how would they significantly impact small businesses and
what can we do to ensure that they are not unfairly burdened?

I have been a strong advocate for including broadband in the universal service
system. Unless small businesses in rural America can take advantage of high-speed,
affordable broadband they will be at a serious competitive disadvantage as compared
with businesses in the urban centers and internationally. By including broadband in the
universal service system we can help ensure that small businesses are positively affected
because the Universal Service Fund can be used to make it more economical for
broadband to be deployed in many places where it is unavailable today. In order to
ensure that small businesses, and contributors generally to the Fund, are not unfairly
burdened, I have proposed that contributions should also be collected from broadband
providers. It seems only fair that if we are to include broadband on the distribution side
of the universal service system that it should also be included on the collection side of the
ledger. In addition, while it would require a legislative fix, including intrastate as well as
interstate revenue when assessing contributions would expand the base for contributions
and be another important tool for evening out the impact of any increase in the Fund.

2.   In March 2004, President Bush spoke about the role broadband plays in the
economic viability of our communities by stating that "we ought to have a universal,
affordable access for broadband technology by the year 2007, and then we ought to
make sure as soon as possible thereafter, consumers have got plenty of choices when
it comes to purchasing the broadband carrier." Despite the President's initiative,
according to the Organization for Economic Cooperation and Development
(OECD), the U.S. has dropped from 4th to 12th among countries in broadband
subscribership since 2000. I'm concerned that our nation is taking steps backwards
rather than improving our nation's broadband accessibility and affordability -
particularly to small businesses in rural parts of the country. Commissioner, it has
been over three years since President Bush made this statement. Why has there been
such a delay in developing and implementing a national broadband policy? How can
we in Congress and the FCC work together to achieve the vitally important policy
underlying this initiative? Wouldn't implementing a national broadband policy

greatly assist small businesses in remaining competitive in this increasingly global economy? In your estimations, how specifically would national broadband policy help small businesses to drive our nation's economy?

I wholeheartedly agree with your concern that the nation is going backwards rather than forwards when it comes to deploying affordable broadband in rural parts of the country. An important reason for the delay in developing a national broadband policy is that we do not have a national commitment to get the job done. Additionally, there needs to be someone looking at the entire playing field and bringing all the piece parts together. This means much more than merely a campaign promise if we are serious about achieving ubiquitous broadband in the country.

As I mentioned in my statement there is a lot that the FCC can do -- collect better data, act as a clearinghouse for broadband ideas, act on our special access rulemaking, and include broadband as part of Universal Service. However, that's not enough. I believe that the delay stems from the lack of a national vision or strategy. The most important thing that can be done would be to put together a national strategy. A good start would be identifying what can the Executive Branch do (USDA, Commerce, Homeland Security - all working together); what is Congress's role in providing tax incentives, providing grants to map where broadband is and is not, and ensuring that public safety gets the benefits of broadband communications; and how should the public and private sector work together to achieve this goal. I am convinced that such a national strategy is essential to our future economy and to the success of small businesses.

3.   The Internet Tax Freedom Act, which is set to expire on November 1, 2007, places a moratorium on the ability of state and local governments to impose: (1) new taxes on Internet access; or (2) any multiple or discriminatory taxes on electronic commerce. I am greatly concerned that allowing the moratorium to expire would be devastating to small businesses. For example, increasing the cost of Internet access to home-based consumers would reduce both consumer demand and telecommunication companies' incentive to build rural broadband networks. Without broadband, rural small businesses would be unable to access high-speed Internet and truly compete in the global marketplace. What would the adverse impacts be on small businesses if we begin to allow state and local governments to tax Internet access? How would this potential expiration inhibit the ability and feasibility of small businesses to do business over the Internet?

As you know, subsequent to the hearing, Congress extended the internet tax moratorium for seven years. Because taxes on internet access are the prerogative of Congress and not the FCC, I have not previously taken a position on the legislation enacting the moratorium. I do believe that any policies that promote the deployment of high value broadband to small businesses are worthy of serious consideration by the Congress.

4.   In your testimony you note that the most recent broadband penetration rankings show that the U.S. is ranked between 11th and 25th in the world. At the same time, other countries who are leading the way in broadband penetration have

much higher population densities. When compared to these countries, the U.S. is made up of unique geography and has a more spread out population density with 32 people per square kilometer. **Is population density a key factor in broadband deployment and if so, is the FCC considering proper policies to suit our unique geography? Why or why not?**

Population density is one of many factors that must be considered when it comes to a strategy for broadband deployment. There are many broadband rankings that have the United States ranked between 11th and 25th. What's worse is that we're headed in the wrong direction. This is not just an embarrassment for a country that prides itself as the technology leader, but it also comes at to high a cost for our economy and for those communities without access broadband. Some apologists for America's dismal performance in these studies argue that those countries that rank higher than the United States live in high rises and more densely populated cities that make it easier to provide access to broadband. In fact, many of the countries ranked ahead of the United States are less densely populated.

So I believe that other factors like income, price, market competition and public policy play as much, if not a larger role than geography in determining broadband availability and adoption. The FCC is finally considering ways to improve its data collection for broadband so that we can better evaluate those areas, particularly rural areas, that don't have access to broadband. Congress is also considering legislation that will assist in better mapping broadband deployment. These are important first steps to developing better policies that specifically address the lack of broadband in less densely populated areas.

5.   **The argument has been made by the FCC and others that we don't need any consumer safeguards for Internet users - including net neutrality- because there is robust competition in the broadband market. However, most people in any state are lucky if they have more than one choice for broadband provider. Do you consider the broadband marketplace to be highly competitive? If the market place is competitive, why do we continue to experience a lack of broadband penetration, specifically in rural areas? I hear a lot of talk about how wireless broadband from a cell phone tower is a viable alternative for DSL or cable modem. Do you believe a wireless connection is a capable substitute for DSL or cable Internet?**

I don't think the broadband marketplace is highly competitive. Today, 96% of the residential broadband market is served by either cable or a Bell phone company. And in a recent GAO study, examination of the special access market (for bulk telephone and broadband services) reveals that around 94% of commercial buildings are served exclusively by the incumbent telephone company. I would not call this competitive. Indeed, the lack of competitive choices is one of the reasons that there is a lack of broadband penetration in the country. I believe that competition means more than just competition between the cable and the phone provider. Our small businesses deserve better than they are getting.

While wireless broadband may become a viable substitute for DSL and cable broadband, these technologies today generally have different characteristics (e.g., speed) that make them attractive for different purposes. I believe that there is a place for all technologies – DSL, cable, and wireless when it comes to bringing broadband to rural communities.

6.   Small businesses are the backbone of our nation's economy representing 99.7 percent of all employer firms and generate three-quarters of net new jobs annually over the last decade. Clearly, small businesses cannot be ignored when it comes to expanding communications technology in our economy. With the critical role small businesses play in driving our economy, what specifically is the FCC doing to reach out to this vital sector? Can you provide some examples of successful small business outreach by the FCC?

The FCC has an Office of Communications Business Opportunities (OCBO) which is supposed to be the principal advisor to the Chairman and the Commissioners on policies affecting small, women- and minority-owned communications businesses. Unfortunately, it appears that its role has been diminished and some even believe that it has been relegated, to a large degree, to a document reduction role. It can and should be doing so much more. Among the things the FCC should be doing more of is working with the Small Business Administration to find proactive ways to increase opportunities for small businesses. When I was Assistant Secretary of Commerce for Trade in the Clinton Administration, we looked for ways to expand exports not just for the Boeings of the world but especially for small businesses. We always were mindful of the fact that small businesses are the engine of the American economy. We don't do enough of that at the FCC.

7.   In the November 2006, the Government Accountability Office (GAO) Report, "FCC Needs to Improve its Ability to Monitor and Determine the Extent of Competition in Dedicated Access Services," concluded that without more complete and reliable data, the FCC is unable to determine whether its deregulatory policies are achieving their goals. However, the FCC responded that gathering additional data would be too costly and burdensome. Do you agree with this assessment? At the hearing, Mr. Mefford of Connect Kentucky testified about the effectiveness of a broadband mapping program. Would mapping the country be an effective way to gather broadband penetration data? What are some problems, if any, with broadband mapping, specifically in relation to small businesses? Do you agree that the Connect Kentucky model would be a preferable alternative to the FCC's current data collection methods? Please provide your thoughts on how you would like to see the FCC change its broadband data collection methods? Do you agree that the FCC's current zip code collection method is flawed? Why or why not?

Broadband data collection has been one of my biggest priorities since I came to the Commission. You can't set good policy if you can't measure the problem. And it's almost as if the FCC's current data collection methodology is designed to deny that there is a problem in the first place. We still say that broadband is 200 kbps and that if anyone in a ZIP Code has broadband then everyone that ZIP Code does. We need to start developing credible measures of speed and deployment, as well as to gather data on price and international experiences with broadband.

With regard to special access, while I believe the FCC could gather additional data, I don't believe additional fact-finding is necessary to begin revising our policies to better reflect the marketplace. In addition to the GAO report, the FCC has had a proceeding open since 2002, reviewed special access in the context of three Bell mergers, and recently requested more data from all parties. In my view, the FCC is long overdue in completing a review of its policies governing special access.

Broadband mapping is certainly an important tool. Connect Kentucky has seen significant success in broadband deployment as a result of its mapping efforts. As for small businesses, I think the answer is to start gathering data on speed tiers. We need to recognize that different users--like a home, a home office, a company with 10 employees, and one with 250--will need very different types of broadband products. Our goal should be to get a real snapshot of what different products are available at given representative locations, in order to understand the choices that real businesses--of all sizes--actually face. Once we understand where we are today, we can start developing policies about how to build a stronger future--one that is more hospitable to small business.

8.   Under the Regulatory Flexibility Act (RFA), Federal agencies are required to consider the impact of their regulatory proposals on small entities and to analyze alternatives that would minimize this impact. The Small Business Administration's Office of Advocacy, the "regulatory watchdog" for small businesses, has expressed repeated concern with the FCC's adherence to its obligations under the RFA. According to the Office of Advocacy's 2006 Report on the Regulatory Flexibility Act, "one of the reasons the FCC has not had consistent compliance with the RFA is its tendency to issue vague proposed rulemakings. Without specific rules, the [FCC] cannot accurately estimate the impacts and assess alternatives to the rule, nor can small businesses comment meaningfully. The FCC has continually rejected Advocacy's recommendations to propose more concrete rules." Do you agree with the Advocacy's finding that the FCC fails to properly comply with the RFA? Why or why not? What steps will the FCC take to address the Office of Advocacy's concerns, specifically with emergent broadband issues like forbearance and copper retirement?

As discussed in more detail in response to question 6, I believe the FCC can and should be doing more when it comes to developing policies specifically to address small business concerns. With regard to issues like forbearance and copper retirement, the impact of these policies and procedures, particularly as they impact competitive telecommunications carriers who often serve small business and are often small businesses themselves, is of the utmost importance. Unfortunately, too often the Commission does not sufficiently take these concerns into account.

9.   This year, Verizon and FairPoint Communications announced that FairPoint Communications will acquire Verizon's properties in Maine, Vermont, and New Hampshire. This $2.7 billion proposed transaction will move FairPoint from the 17th largest telecommunications company to the 8th largest. In my home state, this

transaction makes FairPoint the phone company in almost 90 percent of households. While I am encouraged by the promises that FairPoint has made about increased jobs and accelerated broadband deployment, many Verizon employees in my state have expressed concerns. At the top of the list, they fear that FairPoint will not be strong enough financially to keep their promises after a merger is complete. I understand that the FCC must approve the transfer of certain licenses before the merger is complete in order to be sure the transaction is in the public interest. FairPoint has said that if the acquisition were approved, it would add jobs and improve Internet with $16.1 million in service upgrades. Do you have any reason to believe that FairPoint is incapable of fulfilling these promises? How could this potential merger impact small businesses, specifically in rural areas?

As you know, the transaction is currently pending before the FCC so I am limited in my ability to discuss the specifics of the transaction. I have also heard that there are serious concerns about the financial health of the company after the transaction and its ability to keep its promises to its workers and to deploy broadband in rural areas. These are issues that I will examine closely in considering this transaction.

10.   As you well know, in addition to being an economic driver, broadband also enables revolutionary cultural, political, and educational exchange. Many of the pending Universal Service Fund (USF) reform proposals in the House and Senate include provisions that expand the universal service distribution to cover broadband expenses. For example, the "Universal Service for All Americans Act," which I have cosponsored, would create a $500 million fund within the USF to help support deployment in areas currently unserved by broadband service. I believe that the USF should be adapted to help support some level of broadband expenses, but I fear that if done improperly, the USF could skyrocket to a size that consumers and service providers that pay into the fund will be unwilling to bear. Please discuss your opinion on the legislation that has been introduced in the Senate. In your esteemed opinion, what would be an appropriate way to target support for broadband deployment in a way that does not cause an undue expansion of the fund? What benefits would broadband deployment have on the software and hardware industries, small business and other areas of commerce? What would be the socioeconomic benefits of broadband deployment to under served areas?

I think that broadband is essential to the mission of universal service for the 21st century just as plain old telephone service was the mission of USF in the 20th century. Broadband will certainly be the economic driver in so many areas of the economy. It also is critical in so many other areas such as education, health care, and job opportunities. Creating a broadband fund, as the "Universal Service for All Americans Act" would do, is certainly an important step in bringing broadband out to these areas. I'm pleased that the Joint Board recently agreed with me on a bipartisan basis and now supports broadband as part of the system. Specifically, the Joint Board recommended including broadband in universal service by proposing a broadband fund. Unfortunately, the Joint Board only recommended including $300 million for the broadband fund, which is less than the legislation proposes and is far less expansive than what is necessary to address the most critical infrastructure challenge of our time – bringing broadband to all Americans.

In terms of targeting the support, I believe that support should first go where it can do the greatest good. Many rural companies are doing a really good job of getting broadband out to most of their territory as best as I can tell. But we repeatedly hear from companies that they are bringing broadband out to just 80 or 85% of their service area and that it's just not economic to go to the most remote rural areas. Therefore, ensuring that broadband goes to unserved or largely underserved areas is critically important.

11.   Footnote 92 of the FCC's Order in the April 21, 2004 Declaratory Ruling whether AT&T's Phone-to-Phone IP Telephony Services are Exempt from Access Charges stated:

> *We note that, pursuant to section 69.5(b) of our rules, access charges are to be assessed on interexchange carriers. 47 C.F.R. 69.5(b). To the extent terminating Local Exchange Carriers seek application of access charges, these charges should be assessed against interexchange carriers and not against any intermediate LECs that may hand off the traffic to the terminating LECs, unless the terms of any relevant contracts or tariffs provide otherwise.*

Despite your statement that Interexchange Carriers were the ones liable for access charges and "intermediate Local Exchange Carriers (LECs)" should not be assessed access charges by Incumbent Local Exchange Carriers (ILECs) I remain concerned that AT&T, Verizon, Qwest, and others continue to send large bills to competitive LECs for access charges the ILECs contend are due. I fear that this is creating uncertainty and litigation, and it forces small competitive LECs to either quit serving Voice Over Internet Protocol (VoIP) companies – even when a VoIP company certifies that the traffic does not meet the specific criteria for application of access charges spelled out in one of the Declaratory Rulings – or spend hundreds of thousands of dollars defending themselves. What can the small competitive LECs that are being hit with big bills do about this? Will the FCC entertain a complaint against the large ILECs, and expedite a decision? Why or why not?

Certainly, the FCC should entertain any complaint against a regulated entity that is not complying with the terms of a Commission *Order*. The Commission also owes affected parties—especially new, competitive entrants—a timely decision. After all, business cannot operate under a question mark.

I also note that this type of situation arises at least in part because the FCC has chosen—over my strong objection—to classify a variety of communications services as Title I "information services." By casting aside a body of Title II case law that has arisen over decades in favor of the uncharted waters of Title I, the Commission has robbed consumers and businesses of the protections and regulatory certainty that they are accustomed to and that they deserve.

12.   During Congressional hearings surrounding the Southwestern Bell Corporation (SBC)-AT&T merger, Dave Dorman (former CEO of AT&T) suggested two contradictory reasons why the SBC-AT&T merger in the public interest. First, he said that SBC and AT&T should be allowed to merge because

there are so many other unaffiliated IP-based communications providers out there that AT&T would not be missed as a competitor. Mr. Dorman also said that AT&T could not survive as an unaffiliated player without its own last mile access facilities. Please provide your thoughts on the following: If AT&T, the largest of all the unaffiliated providers, could not survive without access to its own last-mile facilities, how does AT&T expect the other, small, would-be competitors, to survive? In a world where AT&T and Verizon now run powerful reintegrated telecom businesses, how are small rivals expected to compete?

Since I joined the Commission in 2001, I have been concerned that competition in the telecom industry has suffered as the Regional Bell Operating Companies have continued to consolidate. With AT&T and Verizon owning the vast majority of the last mile facilities in their territories, it has become increasingly difficult for unaffiliated providers to compete. To make matters worse, the FCC in recent years has deregulated these companies in many respects, removing regulations that used to provide some assurance that competitors would have access to monopoly facilities at reasonable prices. As a result, deregulation has made it even more difficult for competitors. In my view, the FCC should be more active in fostering competition by more carefully scrutinizing mergers and considering rules that will promote competition rather than allowing just a few large companies to survive. Consumers deserve more choices and competition than just a few reintegrated telecom businesses can provide.

13.   The Supreme Court ruled in the *Trinko* and *Twombley* cases that antitrust laws arguably do not apply to telecom because of the complex regulatory regime and the FCC's ability to keep monopoly power in check. How does government ensure a competitive marketplace without reliance on either antitrust or a complex regulatory regime?

There are really two principal checks on monopoly power: regulation and the antitrust laws. Ensuring a competitive marketplace without vigorous antitrust enforcement and regulatory oversight is like a boxer fighting with both hands tied behind his back. Based on my consideration of mergers that have been jointly reviewed by the FCC and the Justice Department, it does not appear to me that the latter has aggressively enforced the antitrust laws when it comes to telecommunications. Put this together with the fact the Supreme Court's recent decisions may also have reduced the application of the antitrust laws to telecom, more of the burden falls on the effectiveness of telecom regulations to protect against unlawful monopolies. Unfortunately, too often, the FCC in recent years has not proactively or aggressively recognized this as a vital responsibility. In fact, the reverse appears to be true as we continue along the road of deregulation and the reclassification of services. Taken together, there has been, in my view, far too much consolidation in the communications industries that the FCC oversees.

14.   Admittedly, the new "auction" rules for the 700 "Spectrum" are complex, and include one single band that has a limited "open" requirement, which now Verizon is legally challenging as benefiting "Google" and as inappropriate. But, I am concerned that the FCC failed to analyze the benefits of copying the 802.11 (2.4 gig was allocated to it) economic model. Given that there are 5 available "swaths"

of spectrum to allocate, why can't one of these swaths be given to the smaller carriers? If a true economic study was done on the highest and best use, giving some of the air back to smaller carriers may prove to promote more growth and investment in our society just like given away 802.11 did. Small business wants a way to use technology on their own terms, not a service on how to use the technology on AT&T's terms. Also, if the auction process is delayed due to legal action, would there be any prohibition from amending the process and simply "giving back" one swath to the public like 802.11? At the very least, should AT&T and Verizon be prohibited from bidding until anticompetitive concerns described above are dealt with? Why or why not?

I am a big believer in increasing the amount of unlicensed spectrum, which has been an important source of economic growth, consumer well-being, and opportunity for small businesses. I believe that we should always keep unlicensed spectrum in mind as an option as we make decisions about spectrum.

I also believe that an open access model for licensed spectrum also has an important role to play in ensuring that small businesses and entrepreneurs have access to spectrum and freedom to develop innovative technologies that benefit American consumers. To me, open access means both giving consumers freedom to attach non-harmful devices of their choosing as well as giving wireless entrepreneurs the ability to lease spectrum at wholesale prices.

In this summer's 700 MHz Second Report and Order, the Commission set the ground rules for how some of the most valuable spectrum on earth will be used. I am pleased that my colleagues were willing to implement so-called *Carterfone* provisions for at least a portion of this spectrum. I believe this decision will return some power to consumers and entrepreneurs and limit incumbents' power to extract monopoly or oligopoly rents. The device and application openness principles that we implemented for 22 MHz of the commercial spectrum will mean more choices, better services and lower prices. They will permit entrepreneurs to innovate without asking somebody else for permission—just as the developers of the fax machine, dial-up modem, and Wi-Fi router did.

We also took action to prevent abuse in this band and to give consumers, device manufacturers, and other interested parties a right to seek redress if the C-block licensee seeks to discriminate against them. I believe that this case-by-case approach strikes the appropriate balance between preventing harm to the network and giving teeth to our anti-discrimination mandate. Justice delayed is often justice denied, the old adage says, and that is why I am happy that we adopted a 180-day shot clock for Commission enforcement decisions.

Unfortunately, the Commission did not take the additional step—which I would have preferred—of mandating wholesale access principles for some or all of the 700 MHz spectrum. As I noted in my separate statement, a true open access regime would be of enormous benefit for small business and other entrepreneurs, who currently face substantial obstacles in obtaining spectrum access.

As for the involvement of wireline incumbents in the 700 MHz auction, I have long been on record as having concerns about the fact that the two largest wireless providers in the United States are wholly or partially owned by the two largest wireline providers. I believe that these companies may therefore have diminished incentives to use their wireless holdings to compete with DSL and other wireline broadband products. If the Commission had retained spectrum caps several years ago—as I would have preferred—it could have prevented this and other ills stemming from excessive consolidation in the wireless industry.

15.   As the FCC has begun to impose obligations on emerging providers of IP-based voice services (such as universal service payment obligations), while relieving larger, established DSL access providers of the same obligations, what has the FCC done to ensure that these additional obligations are not an excessive burden on small businesses in compliance with the Regulatory Flexibility Act? Please specifically describe the FCC's requirement efforts in your answer.

In order to ensure small businesses are not unduly burdened by FCC rules, the Commission's Office of Communications Business Opportunities (OCBO) is charged with reviewing every rulemaking to ensure that we are in compliance with the Regulatory Flexibility Act. For example, prior to enacting rules such as imposing Universal Service Fund contributions obligations on VoIP providers, the Commission undertook an Initial Regulatory Flexibility Analysis (IRFA) and then provided the opportunity for public comment on the proposals, including comments on the IRFA. In fact, OCBO reviews every rulemaking to ensure that we are in compliance with the letter of the law.

However, we must recognize that the FCC has created too few new rules designed specifically to *help* small businesses. And maybe most importantly, we must realize that some of the Commission's actions – indeed more than a few – have harmed small businesses. I think the FCC can and should do more. We should go beyond simple compliance. We should ask ourselves hard questions consistent with the spirit and intent of the RFA as well.

TESTIMONY OF THE HONORABLE JOHATHAN STEVEN ADELSTEIN

**Questions for the Record from Senator Olympia J. Snowe
Senate Committee on Small Business and Entrepreneurship
to Commissioner Jonathan S. Adelstein, FCC
"Improving Internet Access to Help Small Business Compete in the Global
Economy"**

1. Question: During the Committee's recent hearing, witnesses on both panels testified about the need for an updated data collection methodology. You testified that the FCC's current data is not adequate and does not give us a real picture of what is happening. How can the FCC or Congress effectively reform our current broadband system before knowing the extent of the problem? In your mind, is creating a national mapping strategy the number one broadband priority? Please explain.

Response: I share your concerns about the Commission's data collection efforts to gauge broadband deployment, access, and affordability. The Commission's efforts thus far fall far short of our obligations under Section 706 of the Act to provide reports on the status of broadband deployment. In its May 2006 report, the Government Accountability Office (GAO) took the FCC to task for the quality of its broadband data. GAO criticized the Commission's ability to analyze who is getting broadband and where it is deployed, observing that the FCC's data "may not provide a highly accurate depiction of deployment of broadband infrastructures for residential service, especially in rural areas." The clear conclusion is that FCC has much work to do to improve the quality and scope of its broadband data, as well as its analysis of the availability of affordable broadband services, if it is to satisfy the Congressional mandate in Section 706 of the Act. If this is not our highest priority, it is clearly must be a first step toward improving our treatment of these issues.

2. Question: In March 2004, President Bush spoke about the role broadband plays in the economic viability of our communities by stating that "we ought to have a universal, affordable access for broadband technology by the year 2007, and then we ought to make sure as soon as possible thereafter, consumers have got plenty of choices when it comes to purchasing the broadband carrier." Despite the President's initiative, according to the Organization for Economic Cooperation and Development (OECD), the U.S. has dropped from 4th to 12th among countries in broadband subscribership since 2000. I'm concerned that our nation is taking steps backwards rather than improving our nation's broadband accessibility and affordability – particularly to small businesses in rural parts of the country. Commissioner, it has been over three years since President Bush made this statement. Why has there been such a delay in developing and implementing a national broadband policy? How can we in Congress and the FCC work together to achieve the vitally important policy underlying this initiative? Wouldn't implementing a national broadband policy assist small businesses in remaining competitive in this increasingly global economy? In your estimations, how specifically would national broadband policy help small businesses to drive our nation's economy?

Response: One of America's central challenges is promoting the widespread deployment of higher-bandwidth broadband facilities to carry the vast array of new innovative services that are transforming virtually every aspect of the way we communicate, and to make sure that these facilities are affordable for consumers. This must be a greater national priority than it is now. An issue of this importance to the economy of our nation and the success of our communities warrants a coherent, cohesive, and comprehensive national broadband strategy.

The U.S. needs a national broadband strategy that seriously addresses our successes and failures, and strives to improve our broadband status. Virtually every other advanced country has implemented a national broadband strategy. Even though we have made strides, I am concerned that the lack of a comprehensive plan is one of the reasons that the U.S. is nevertheless falling further behind our global competitors. As you note, we continue to slip further down the regular rankings of broadband penetration. More troubling, there is growing evidence that citizens of other countries are getting a much greater broadband value, in the form of more megabits for less money. According to the ITU, the digital opportunity afforded to U.S. citizens is not even near the top, it's 21st in the world. This is more than a public relations problem. It's a productivity problem, and our citizens deserve better.

We must engage in a concerted and coordinated effort to restore our place as the world leader in telecommunications by making affordable broadband available to all our citizens. It will mean taking a hard look at our successes and failures, and improving our data collection. A true broadband strategy should incorporate benchmarks, deployment timetables, and measurable thresholds to gauge our progress. It is not enough to rely on poorly-documented conclusions that deployment is reasonable and timely.

We need to set ambitious goals, shooting for real high-bandwidth broadband deployment. We should start by updating our current definition of high-speed of just 200 kbps in one direction to something more akin to what consumers receive in countries with which we compete, speeds that are magnitudes higher than our current definitions. Further, we need much more reliable data than the FCC currently compiles so that we can better ascertain our current problems and develop responsive solutions. Giving consumers reliable information by requiring public reporting of actual broadband speeds by providers would spur better service and enable the free market to function more effectively.

We must also re-double our efforts to encourage broadband development by increasing incentives for investment because we will rely on the private sector as the primary driver of growth. These efforts must take place across technologies so that we not only build on the traditional telephone and cable platforms, but also create opportunities for deployment of fiber-to-the-home, fixed and mobile wireless, broadband over power line, and satellite technologies.

We must also work to promote meaningful competition, as competition is the most effective driver of lower prices and innovation. This is increasingly important to ensure that the U.S. broadband market does not stagnate into a comfortable duopoly, a serious concern given that cable and DSL providers control 96 percent of the residential broadband market.

The Commission must also ensure the vitality of universal service as technology evolves. With voice, video, and data increasingly flowing to homes and businesses over broadband platforms, it will be critical to have ubiquitous high speed networks to carry these services everywhere. This means that universal service must evolve, as Congress intended, to cover broadband services. We must also promote spectrum-based services that can play such an important role spurring both competition and greater availability of these services.

There also is more Congress can do, outside of the purview of the FCC, such as providing adequate funding for Rural Utilities Service broadband loans and grants, and ensuring RUS properly targets those funds; establishing new grant programs supporting public-private partnerships that can identify strategies to spur deployment; providing tax incentives for companies that invest in broadband to underserved areas; devising better depreciation rules for capital investments in targeted telecommunications services; promoting the deployment of high speed Internet access to public housing units and redevelopments projects; investing in basic science research and development to spur further innovation in telecommunications technology; and improving math and science education so that we have the human resources to fuel continued growth, innovation and usage of advanced telecommunications services; and, of course, we need to make sure all of our children have affordable access to their own computers to take full advantage of the many educational opportunities offered by broadband.

3. Question: The Internet Tax Freedom Act, which is set to expire on November 1, 2007 places a moratorium on the ability of state and local governments to: (1) impose new taxes on Internet access; or (2) any multiple or discriminatory taxes on electronic commerce. I am greatly concerned that allowing the moratorium to expire would be devastating to small businesses. For example, increasing the cost of Internet access to home-based consumers would reduce both consumer demand and telecommunications companies' incentives to build rural broadband networks. Without broadband, rural small businesses would be unable to access high-speed Internet and truly compete in the global marketplace. What would the adverse impacts be on small businesses if we begin to allow state and local governments to tax Internet access? How would this potential expiration inhibit the ability and feasibility of small business over the Internet?

Response: While it is my understanding that the Internet Tax Freedom Act of 2007 does not confer any specific jurisdiction to the Federal Communications Commission, I note that this Act became public law on October 31, 2007.[1] According to the Congressional Research Service, the Internet Tax Freedom Act Amendments Act of 2007 amends the Internet Tax Freedom Act to: (1) extend until November 1, 2011, the moratorium on state and local taxation of Internet access and electronic commerce and the exemption from such moratorium for states with previously enacted Internet tax laws; (2) restrict the authority of certain states claiming an exemption from the moratorium under the Internet Tax Nondiscrimination Act of 2004 to impose Internet access taxes after November 1, 2007; (3) expand the definition of "Internet access" to include related communication services (e.g., e-mails and instant messaging) and redefine "telecommunications" to include unregulated non-utility telecommunications (e.g., cable service); and (4) allow a specific exception to the moratorium for certain state business taxes enacted between June 20, 2005, and November 1, 2007, that do not tax Internet access.

---

[1] *See Library of Commerce, THOMAS (http://thomas.loc.gov/cgi-bin/bdquery/z?d110:H803678:@@@L&summ2=m&).*

4. Question: As you well know, in addition to being an economic driver, broadband also enables revolutionary cultural, political, and educational exchange. Many of the pending Universal Service Fund (USF) reform proposals in the House and Senate include provisions that expand the universal service distribution to cover broadband expenses. For example, the "Universal Service for All Americans Act," which I have cosponsored, would create a $500 million fund within the USF to support deployment in areas currently unserved by broadband service. I believe that the USF could skyrocket to a size that consumers and services providers that pay into the fund will be unwilling to bear. Please discuss your opinion on the legislation that has been introduced in the Senate. In your esteemed opinion, what would be an appropriate way to target support for broadband deployment in a way that does not cause an undue expansion of the fund? What benefits would broadband deployment have on the software and hardware industries, small business and other areas of commerce? What would be the socioeconomic benefits of broadband deployment to underserved areas?

Response: Congress and the Commission recognized early on that the economic, social, and public health benefits of the telecommunications network are increased for all subscribers by the addition of each new subscriber. Federal universal service continues to play a vital role in meeting our commitment to connectivity, helping to maintain high levels of telephone penetration, and increasing access for our nation's schools and libraries.

Ensuring the vitality of universal service will be particularly important as technology continues to evolve. As voice, video, and data increasingly flow to homes and businesses over broadband platforms, voice is poised to become just one application over broadband networks. So, in this rapidly-evolving landscape, we must ensure that universal service evolves to promote advanced services, which is a priority that Congress made clear.

I note that the Federal-State Joint Board on Universal Service (Joint Board) recently recommended that the Commission revise its list of services supported by Federal universal service to include broadband Internet access service. The Joint Board recommended that the Commission establish a Broadband Fund, tasked primarily with facilitating construction of facilities for new broadband services to unserved areas. The Joint Board also recognized the effectiveness of the current High Cost Loop Fund in supporting the capital costs of providing broadband-capable loop facilities for rural carriers. I look forward to carefully reviewing the Joint Board's recommendations, and I hope that the Commission will seek comment quickly on these from a broad range of commenters.

Finally, I note that it is important that the Commission conduct its stewardship of universal service with the highest of standards. I have worked hard to preserve and advance the universal service programs as Congress intended, will continue to do so, and look forward to any guidance from Congress regarding this important program.

5. Under the Regulatory Flexibility Act (RFA), Federal agencies are required to consider the impact of their regulatory proposals on small entities and to analyze alternatives that

would minimize this impact. The Small Business Administration's Office of Advocacy, the "regulatory watchdog" for small businesses, has expressed repeated concern with the FCC's adherence to its obligations under the RFA. According to the Office of Advocacy's 2006 Report on the Regulatory Flexibility Act, "one of the reasons the FCC has not had consistent compliance with the RFA is its tendency to issue vague proposed rulemakings. Without specific rules, the [FCC] cannot accurately estimate the impacts and assess alternatives to the rule, nor can small businesses comment meaningfully. The FCC has continually rejected Advocacy's recommendations to propose more concrete rules." Do you agree with the Advocacy's finding that the FCC fails to properly comply with the RFA? Why or why not? What steps will the FCC take to address the Office of Advocacy's concerns, specifically with emergent broadband issues like forbearance and copper retirement?

Response: Small businesses play a critical role in creating jobs and developing new technologies. They also purchase a massive amount of telecommunications services, spending approximately $25 billion each year, according to a recent Wall Street Journal report. So, it is clear that the Commission's decisions are likely to impact the telecommunications services and opportunities available to small businesses.

As you observe, the RFA requires the Commission to analyze the economic impact of draft regulations when there is likely to be a significant economic impact on a substantial number of small entities, and to consider regulatory alternatives that minimize the burden on small entities. So, I am concerned about the Office of Advocacy's findings regarding the FCC's compliance with the Regulatory Flexibility Act (RFA).

With respect to forbearance, I share the Office of Advocacy's concerns. The Commission's recent history on forbearance petitions – including failing to even issue an order addressing the merits of a sweeping petition – is not one to be envied. This approach has cast open the floodgates for industry-filed petitions, inviting parties to make end runs around the Congressional framework for telecommunications services. I have repeatedly urged the Commission to adopt procedural rules for forbearance petitions, such as requiring parties to include in their original petitions detailed information about the services subject to the petition and a detailed analysis of how such proposals satisfy the statutory test. Procedural rules can provide transparency and predictability to all interested participants and can restore confidence in Commission processes.

Similarly, I agree with the Office of Advocacy that the Commission should look closely at our policies regarding the retirement of copper facilities. Two currently pending petitions ask the Commission to investigate whether the retirement of copper facilities would lessen the redundant capabilities available for consumers, including federally owned and leased buildings. These petitions argue that copper loop and subloop retirement eliminate network alternatives that might otherwise prove essential for network redundancy in the event of a homeland security crisis, natural disaster, or the recovery period after such events. The Commission has recognized the importance of redundant communications in several contexts. Indeed, the Independent Panel Reviewing the Impact of Hurricane Katrina on Communications Networks found that failure of redundant pathways for communications traffic was one of three main problems that caused the majority of communications network interruptions. The Commission has sought comment on these petitions, and I look forward to reviewing the record developed in response.

6. This year, Verizon and FairPoint Communications announced that FairPoint Communications will acquire Verizon's properties in Maine, Vermont, and New Hampshire. This $2.7 billion proposed transaction will move FairPoint from the 17th largest telecommunications company to the 8th largest. In my home state, this transaction makes FairPoint the phone company in almost 90 percent of households. While I am encouraged by the promises that FairPoint has made about increased jobs and accelerated broadband deployment, many Verizon employees in my home state have expressed concerns. At the top of the list, they fear that FairPoint will not be strong enough financially to keep their promises after a merger is complete. I understand that the FCC must approve the transfer of certain licenses before the merger is complete in order to be sure that the merger is in the public interest. FairPoint has said that if the acquisition were approved, it would add jobs and improve Internet with $16.1 million in service upgrades. Do you have any reason to believe that FairPoint is incapable of fulfilling these promises? How could this potential merger impact small businesses, specifically in rural areas?

Response: Verizon Communications and FairPoint Communications filed, on January 31, 2007, a series of applications pursuant to Section 214 and 310(d) of the Act seeking Commission approval to transfer control of Verizon's local exchange assets in New Hampshire, Maine, and Vermont from Verizon to FairPoint. The Commission's obligation under Sections 214 and 310(d) is to determine whether the proposed transaction will serve the public interest, convenience, and necessity.

These applications are currently pending before the Commission and I am currently reviewing the record. The Applicants have, as you note, asserted that this transaction would produce numerous public interest benefits, including enhanced service quality, increased capital expenditures, accelerated broadband deployment, and the creation of new jobs in the region. I have also heard concerns from commenters, including the Communications Workers of America (CWA) and the International Brotherhood of Electrical Workers (IBEW), that the public interest would not be served by this transaction due. CWA and IBEW argue, in particular, that FairPoint lacks the financial resources, operational capacity, and experience. According to precedent, the Commission considers in its public interest inquiry whether the applicant for a license has the requisite "citizenship, character, financial, technical, and other qualifications." Given the significant and growing role of telecommunications services in the health of our communities and the potential impact on the citizens and businesses of Maine, Vermont, and New Hampshire, I agree that this is a momentous decision and will carefully consider these issues.

7. Small businesses are the backbone of our nation's economy representing 99.7 percent of all employer firms and generate three-quarters of net new jobs annually over the last decade. Clearly, small businesses cannot be ignored when it comes to expanding communications technology. With the critical role small businesses play in driving our

economy, what specifically is the FCC doing to reach out to this vital sector? Can you provide some examples of small business outreach by the FCC?

Response: The Commission created the Office of Communications Business Opportunities (OCBO) in 1994 to promote business opportunities for entrepreneurs and other small businesses, including minority- and women-owned businesses. OCBO describes outreach among its activities, including:

> "The Office of Communications Business Opportunities (OCBO) promotes telecommunications business opportunities for small, minority-owned, and women-owned businesses. To this end, OCBO works with entrepreneurs, industry, public interest organizations, individuals, and others to provide information about FCC policies, increase ownership and employment opportunities, foster a diversity of voices and viewpoints over the airwaves, and encourage participation in FCC proceedings. OCBO also mails information on Commission notices and new service opportunities to those within our database of over 3,000 small, minority-owned, and women-owned businesses and other interested entities. OCBO periodically co-hosts auction seminars to inform the public about new licensing opportunities, as well as seminars concerning new technologies and business opportunities utilizing unlicensed spectrum. To assist small businesses in understanding and complying with the FCC's rules, the OCBO web site provides a comprehensive list of small business compliance guides."

See http://www.fcc.gov/ocbo/.

I note that Congress has directed the Commission, in Section 257 of the Act, to identify and report on how it eliminated market entry barriers for entrepreneurs and other small businesses. The purpose of our exercise is clear from the statute – to promote the policies and purposes of the Communications Act "favoring diversity of media voices, vigorous economic competition, technological advancement, and the promotion of the public interest, convenience, and necessity."

I share a strong commitment to these Congressional goals. Entrepreneurs and small businesses play a crucial role in communications industries, from providing service in rural and underserved areas, to encouraging innovation and niche operations, to bringing a unique and diverse voice to the public airwaves, and countless other examples. As noted in my prior response, small businesses play a critical role in creating jobs and developing new technologies. They also purchase a massive amount of telecommunications services, spending approximately $25 billion each year.

The FCC's Section 257 reports reveal that the Commission has not done good job in using the Office of Communications Business Opportunities (OCBO) as its "principal small business policy advisor." During the three-year period covered in the most recent report, the Commission has failed to charge OCBO - independently or in conjunction with a Bureau or Office - with developing or launching any significant policies, plans or programs to further the concerns of small businesses. As a consequence of the Commission's misguided priorities, OCBO – which was formed specifically to address the concerns of small businesses and has very talented staff with subject matter expertise – has not played a meaningful role in the policy development process at the Commission. The Commission must continue to expand its efforts to address to promote opportunities for small businesses, including minority- and women-owned businesses.

8. Question: The argument has been made by the FCC and others that we don't need any consumer safeguards for Internet users – including net neutrality – because there is robust competition in the broadband market. However, most people in any state are lucky if they have more than one choice for broadband provider. Do you consider the broadband marketplace to be highly competitive? If the marketplace is competitive, why do we continue to experience a lack of broadband penetration, specifically in rural areas? I hear a lot of talk about how wireless broadband from a cell phone tower is a viable alternative for DSL or cable modem. Do you believe a wireless connection is a capable substitute for DSL or cable Internet?

Response: It is difficult to assess the relative competitiveness of the current broadband services market because of the lack of sufficient data collected at the FCC and because the industry is changing so dramatically. Unfortunately, the Commission's current efforts to gauge broadband deployment, competition, and affordability fall short. In a May 2006 report, the Government Accountability Office (GAO) took the FCC to task for the quality of its broadband data. GAO criticized the Commission's ability to analyze who is getting broadband and where it is deployed, observing that the FCC's data "may not provide a highly accurate depiction of deployment of broadband infrastructures for residential service, especially in rural areas." Similarly, GAO observed that the number of providers reported in a Zip Code overstates the level of competition to individual households. One clear conclusion from the GAO's report is that the Commission must explore ways to develop greater granularity in its assessment and analysis of broadband availability, whether through statistical sampling, Census Bureau surveys, or other means.

The data available suggest that, even though we have made strides with broadband deployment, we must continue to promote meaningful competition. Competition is the most effective driver of lower prices and innovation, so I am concerned that cable and DSL providers control 96 percent of the residential broadband market. We must be vigilant to ensure that the U.S. broadband market does not stagnate into a comfortable duopoly. I also agree with your observation that different Internet connections offer different capabilities. The FCC's low definition of broadband captures services that do not allow users the full capabilities of truly high speed Internet connections comparable to what citizens in other countries receive.

Finally, I believe that consumers must be at the top of our list, not the bottom, as we move into the broadband era. Through the Communications Act, Congress codified a broad set of consumer protection obligations for telecommunications services that the FCC has now side-stepped with its current approach to broadband services. It is regrettable that, two years after exercising the blunt instrument of reclassification, the Commission has not significantly advanced the discussion of safeguards for broadband consumers, even though we have an open docket concerning Consumer Protection in the Broadband Age. The Commission must do more to assess the experiences and expectations of broadband consumers, who deserve our attention.

This is not to suggest that we regulate reflexively or append legacy approaches where they do not belong. It is imperative, however, that the FCC not remain silent, allowing consumers to push forward into the broadband age without taking stock of consumers' experiences and expectations, much less leaving them in a vortex of undefined roles and safeguards. Having heard from an extraordinary number of consumers who are concerned about the future of the Internet, the Commission could do more to engage the public in a dialogue. For example, the Commission has issued a five-page inquiry on broadband industry practices, but we should be actively engaging consumers about their practices, expectations, and opportunities, particularly as Internet users increasingly become producers, not just consumers, of content. In all these efforts, we must also recognize that time is critical, so that we do not continue to leave consumers in legal limbo.

9. Commissioner Adelstein, during your testimony before this Committee, you stated that small businesses are "starved for telecommunications choice," and many small businesses have only one choice of provider for broadband services, leading to higher prices. Additionally, you note that the Small Business Administration's Office of Advocacy finds "the combination of high prices and few alternatives creates an insurmountable burden to small carriers trying to conduct business in the telecommunications market." What role can the FCC play in increasing competition among service providers and ultimately lowering prices?

Response: Unfortunately, the FCC collects little reliable data about extent of broadband services available to small businesses in the U.S., or the more general state of competition among providers of telecommunications services for businesses. In a report released at the end of last year, the U.S. Government Accountability Office (GAO) recommended that the Commission collect additional data to monitor competition and to assess customer choice through, for example, price indices and availability of competitive alternatives. GAO found that "without more complete and reliable measures of competition, [the] FCC is unable to determine whether its deregulatory policies are achieving their goals."

As noted in my testimony, I believe there are encouraging signs that small businesses are integrating new services and features into their business plans. Businesses of all sizes are increasingly tapping into broadband to reduce costs, increase productivity, and improve efficiency. Broadband is connecting entrepreneurs to millions of new distant potential customers, creating opportunities for telecommuting, and giving businesses new tools to increase productivity. Much of the economic growth we have experienced in the last decade is attributable to productivity increases that have arisen from advances in technology, particularly in telecommunications. These new connections increase the efficiency of existing business and create new jobs by allowing new businesses to emerge, and spur new developments such as remote business locations and call centers.

Yet, the picture is far less clear with respect to telecommunications choice for small business. Many small businesses have only one choice of provider for broadband services, which deprives them of innovative alternatives and can result in higher prices. Even where there are competitive options, alternative providers rely heavily on inputs from incumbent. This highlights the importance of pro-competitive policies. Congress directed the Commission in the Telecommunications Act of 1996 to take both pro-competitive and deregulatory actions. In recent years, the Commission has fallen short in its efforts to balance these goals in its handling of industry-filed forbearance petitions. I believe that the Commission must do a much better job in its reconciliation of these twin goals, for example, by adopting procedural rules governing forbearance petitions.

Finally, I note that Section 257 of the Act mandates that, every three years, the Commission review and report on efforts to identify and eliminate regulatory barriers to market entry in the provision and ownership of telecommunications services and information services, and on proposals to eliminate statutory barriers to market entry by those entities. Although I dissented in part from our most recent Report, I will continue to encourage the Commission to improve its efforts to promote opportunities for small businesses, particularly for women and minorities.

11. During Congressional hearings surrounding the Southwestern Bell Corporation (SBC)-AT&T merger, Dave Dorman (former CEO of AT&T) suggested two contradictory reasons why the SBC-AT&T merger is in the public interest. To paraphrase: First, he said that SBC and AT&T should be allowed to merge because there are so many other unaffiliated IP-based communications providers out there that AT&T would not be missed as a competitor. Mr. Dorman also said that AT&T could not survive as an unaffiliated player without its own last mile access facilities. Please provide your thoughts on the following: If AT&T, the largest of all the unaffiliated providers, could not survive without access to its own last mile facilities, how does AT&T expect the other, smaller, would-be competitors to survive? In a world where AT&T and Verizon now run powerful reintegrated telecom businesses, how are small rivals expected to compete?

Response: I share your concern about concentration in the telecommunications marketplace. In my statement to the SBC-AT&T merger, I stated:

> I am concerned about the potential harms of these mergers. AT&T and MCI are, without question, two of the leading providers of competitive choice across the country, and these combinations will, by any measure, create more concentration in markets that are already highly concentrated. We must be particularly careful where a proposed merger would lead to less competition rather than more, so I give these concerns great weight.

> Based on my weighing of these potential benefits and harms, I could not support these mergers in the absence of reasonable conditions. Without conditions, there is a real possibility that these combinations would increase rates for both residential and business consumers and put at risk the continued existence of the open and robust Internet. So, my support here is based on the Applicants' offers to comply with a minimum set of

conditions that will help promote consumer choice and the development competitive alternatives. Indeed, I would have preferred additional and more rigorous safeguards beyond those set forth in these Orders.[2]

Recent mergers in the telecommunications marketplace have been historic in scope, but they are also part of a larger industry restructuring that is quickly changing the landscape for consumers of telephone, Internet and video services. The opportunities from these technologies are greater than ever, but so is the penalty for those left without options. The Commission must continue to do all it can to ensure that consumers enjoy the benefit of competition, and the innovations and lower prices which it brings.

12. The Supreme Court ruled in *Trinko* and *Trombley* cases that antitrust laws arguably do not apply to telecom because of the complex regulatory regime and the FCC's ability to keep monopoly power in check. How does the government ensure a competitive marketplace without reliance on either antitrust or a complex regulatory regime?

Response: As you note, both the Communications Act, as amended by the Telecommunications Act of 1996 (1996 Act), and antitrust law play an important role in ensuring a competitive telecommunications marketplace. Regarding the interaction between these two tools, the Commission observed in 1996 when adopting rules to implement the Telecommunications Act of 1996 that "[n]othing in sections 251 and 252 or our implementing regulations is intended to limit the ability of persons to seek relief under the antitrust laws, other statutes, or common law."[3]

Since that time, the Supreme Court has opined on the effect of the 1996 Act upon the application of traditional antitrust principles. *See Verizon Communications Inc. v. Law Offices of Curtis V. Trinko, LLP*, 540 U.S. 398, 402, 124 S.Ct. 872. The Court stated that "just as the 1996 Act preserves claims that satisfy existing antitrust standards, it does not create new claims that go beyond existing antitrust standards." *Id. at* 872.

While Congress has the power to further clarify the relationship between these two important tools, it is Commission's responsibility to enforce the statutory obligations within our jurisdiction. As the Commission noted back in 1996,[4] there are a wide variety of tools available to do so and I remain committed to promoting competition in the telecommunications markets.

---

[2] See Statement of Commissioner Jonathan S. Adelstein, available at
http://hraunfoss.fcc.gov/edocs_public/attachmatch/FCC-05-184A1.pdf.

[3] *Implementation of the Local Competition Provisions in the Telecommunications Act of 1996*, CC Docket Nos. 96-98, 95-185, Report and Order, 11 FCC Rcd 15499 (1996) (*Local Competition Order*) (subsequent history omitted).

[4] *See Local Competition Order* at para. 129 ("[I]n appropriate circumstances, the Commission could institute an inquiry on its own motion, 47 U.S.C. § 403, initiate a forfeiture proceeding, 47 U.S.C. § 503(b), initiate a cease-and-desist proceeding, 47 U.S.C. § 312(b), or in extreme cases, consider initiating a revocation proceeding for violators with radio licenses, 47 U.S.C. § 312(a), or referring violations to the Department of Justice for possible criminal prosecution under 47 U.S.C. § 501, 502 & 503(a).").

13. Admittedly, the new "auction" rules for the 700 "Spectrum" are complex, and include one single band that has a limited "open" requirement, which now Verizon is legally challenging as benefiting "Google" and as inappropriate. But, I am concerned that the FCC failed to analyze the benefits of copying the 802.11 (2.4 was allocated to it) economic model. Given that there are 5 available "swaths" of spectrum to allocate, why can't one of these swaths be given to the smaller carriers? If a true economic study was done on the highest and best use, giving some of the air back to smaller carriers may prove to promote more growth and investment in our society just like giving away 802.11 did. Small business wants a way to use technology on their own terms, not a service on how to use the technology on AT&T's terms. Also, if the auction process is delayed due to legal action would there be any prohibition from amending the process and simply "giving back" one swath to the public like 802.11? At the very least, should AT&T and Verizon be prohibited from bidding until anticompetitive concerns described above are dealt with? Why or why not?

Response: I have previously advocated the use of unlicensed spectrum as an intriguing avenue for many underserved communities given that unlicensed spectrum is free and, in most rural areas, lightly used. The same holds true for small carriers. Unlicensed spectrum can be accessed immediately, and the equipment is relatively cheap because it is so widely available. I have also pressed for the inclusion of smaller blocks of licenses to offer opportunities for a broad variety of licenses consistent with our statutory obligation under section 309(j). While the FCC's recently adopted rules for the 700 MHz spectrum reflect a compromise among many different competing interests, I am hopeful that there will be opportunities for a diverse group of licensees in the 700 MHz auction and that our more aggressive build-out requirements will benefit consumers across the county.

14. As the FCC has begun to impose obligations on emerging providers of IP-based voice services (such as universal service payment obligations), while relieving larger, established Digital Subscriber Line access providers of these same obligations, what has the FCC done to ensure that these additional obligations are not an excessive burden on small businesses in compliance with the Regulatory Flexibility Act? Please specifically describe the FCC's requirement efforts in your answer.

Response: All indicators suggest that the IP-based services, like VoIP, are rapidly becoming the building blocks for the future of telecommunications. These services promise a new era of consumer choice, and consumers are rapidly adopting these services. The migration to VoIP services also raises questions for consumers, providers, and, in particular, for the Commission.

The Commission is charged under the Communications Act with ensuring that the goals set out by Congress are fulfilled, and the Commission has now issued a series of orders addressing the regulatory obligations applicable to interconnected VoIP providers. As we do so, we must also take into account the impact of our actions on small businesses, including small VoIP providers. The RFA requires the Commission to analyze the economic impact of draft regulations when there is likely to be a significant economic impact on a substantial number of small entities, and to consider regulatory alternatives that minimize the burden on small entities.

Forging the right regulatory scheme for interconnected VoIP services is a critical task. I agree that the Commission could do more to reach out to small VoIP providers when considering changes to our regulatory policies and I will continue to encourage the Commission make decisions in manner that promotes the deployment of new technologies and minimizes the economic impact of our regulatory actions on small providers.

15. Footnote 92 of the FCC's Order in the April 21, 2004 Declaratory Ruling whether AT&T's Phone-to-Phone IP Telephony Services are Exempt from Access Charges stated:

> **We note that, pursuant to section 69.5(b) of our rules, access charges are to be assessed on interexchange carriers. 47 C.F.R. § 69.5(b). To the extent that terminating local exchange carriers seek application of access charges, these charges should be assessed against interexchange carriers and not against any intermediate LECs that may hand off the traffic to the terminating LECs, unless the terms of any relevant contracts or tariffs provide otherwise.**

Despite your statement that IntereXchange Carrier were the ones liable for access charges and "intermediate Local Exchange Carriers (LECs)" should not be assessed access charges by Incumbent Local Exchange Carriers (ILECs) I remain concerned that AT&T, Verizon, Qwest and others continue to send large bills to competitive LECs for access charges the ILECs contend are due. I fear that this is creating uncertainty and litigation, and it forces small competitive LECs to either quit serving Voice Over Internet Protocol (VoIP) companies – even when a VoIP company certifies that the traffic does not meet the specific criteria for application of access charges spelled out in one of the Declaratory Rulings – or spend hundreds of thousands of dollars defending themselves. What can the small competitive LECs that are being hit with big bills do about this? Will the FCC entertain a complaint against the large ILECs, and expedite a decision? Why or why not?

Response: At the time of the AT&T decision, I noted that the one point of unanimity in our record was the desire for a Commission decision. While some parties have asked us to go further and address broader issues regarding intercarrier compensation and regulation of VoIP services, the Commission concluded that delay in answering the question at hand would serve only to create instability for the long distance industry and to increase the rapidly-growing stakes for each side.

Three years since that decision, the Commission still has before it numerous intercarrier compensation issues, and I continue to have serious concerns about the sustainability of the current system. The current system relies on distinctions between different types of carriers and services. In addition, many developments since the adoption of the current rules – such as the rise of VoIP, and new service offerings, such as flat-rate calling plans – have challenged our traditional distinctions. Uncertainty about the application of the current rules has resulted in calls for reform from a wide diversity of interests, including state policymakers, consumer groups, incumbent and competitive local wireline carriers, wireless carriers, long distance carriers, VoIP providers, and others. It is also highly foreseeable that, in the absence of greater clarity from the Commission on the applicability of its rules, carriers will seek to resolve

disputes, whether at the Commission or elsewhere. I will work with my colleagues to address these issues as expeditiously as possible.

SCOTT, BEN
TESTIMONY, PREPARED STATEMENT

Senator Olympia J. Snowe
Witness Questions
SBC Committee Hearing
September 26, 2007

QUESTIONS
Ben Scott

1. **One argument made by the FCC and others is that we don't need any consumer safeguards for Internet users – including "net neutrality" –because there is robust competition in the broadband market. However, most people in any state are lucky if they have more than one choice for broadband provider. Do you consider the broadband marketplace to be highly competitive? If the marketplace is competitive, why do we continue to experience a lack of broadband penetration, specifically in rural areas? I hear a lot of talk about how wireless broadband from a cell phone tower is a viable alternative for DSL or cable modem. Do you believe a wireless connection is a capable substitute for DSL or cable Internet?**

The broadband market is simply not competitive. No amount of rhetoric from incumbents to the contrary can dispute the plain facts. There are two dominant technologies in the market—DSL and cable modem. These services are typically provided by regional monopolies over each technology—one cable company and one phone company. Therefore, each community in America has, at best, a duopoly marketplace for broadband access. I know of no economic standard by which two market players constitutes a competitive marketplace.

According to the most recent FCC data (submitted by the carriers), 93% of residential broadband lines are either cable modem or DSL. This figure is very likely an undercount of actual market dominance in the 97-98% range. This is because the 4% of connections that are tabulated as mobile wireless are often duplicate connections. That is, consumers who own Blackberries (mobile wireless connections) typically also have a home broadband connection. The mobile device is not a substitute for DSL or cable modem service, it is a complement.

There is simply no way to judge this market a competitive one, much less a highly competitive one. This lack of competition causes a number of problems. In particular, there is little incentive to rapidly increase the speed and lower the cost of broadband (which is what has happened in the competitive broadband markets overseas, as opposed to at home, where improvements have been slow and incremental). This makes the product less valuable to consumers and tempers the penetration rate. In rural areas, we have this affect added on top of significant gaps in the incumbent service territories that leave approximately 9% of US households without a wireline broadband provider. Further, according to FCC data, only 79% of telephone lines are DSL capable. Though FCC data is not granular enough to demonstrate it conclusively, it is likely that the remaining 21% are disproportionately in rural areas.

As to the viability of wireless broadband as a substitute to DSL and cable modem, it is not yet, and it does not appear likely to be in the near future. Currently, the wireless and wireline broadband products are in completely different product markets. They are not comparable in either performance or price—wireless broadband is often twice the cost for half the speed in exchange for mobility. They are not substitutable services; and they are certainly not direct competitors. Though the FCC has never bothered to count, it seems obvious that subscribers to mobile broadband devices (such as Blackberries) have not cancelled their wireline broadband service as a result. The wireless product is a complementary product for which the consumer pays extra. Most consumers do not use mobile wireless broadband on cell phones for the same purposes as a residential broadband connection.

While it is true that mobile broadband subscriptions are increasingly rapidly, the vast majority are business users (88.7% according to the FCC), not residential users. What's more, the three largest mobile data carriers are AT&T, Verizon and Sprint. Two of these three carriers are also ILECs. They are the number one (AT&T) and number three (Verizon) most subscribed to broadband Internet service providers, and these carriers are also the top two DSL providers in the United States. Sprint has a joint venture with cable operators. These affiliations with incumbent wireline providers present an incentive for these companies NOT to use their wireless products to cannibalize their wireline broadband market.

2. **The U.S. is made up of a unique geography and population density when compared to other countries. With 32 people per square kilometer the U.S. ranks 12th in population density and 15th in terms of broadband penetration. However, other countries who are leading the way in broadband penetration have much higher population densities. For example, Korea is ranked 2nd in broadband deployment and has a population density of 481 people per square kilometer. Is population density a key factor in broadband deployment, and if so, are we implementing proper policies to suit our unique geography? Why or why not?**

Apologists for the poor U.S. broadband numbers are quick to attribute the low penetration level to this country's relatively low population density. It makes a certain intuitive sense, so we decided to study the question empirically using an econometric model to measure the impact of population density on the relative performance of the US next to other nations. We found that for the 30 nations of the OECD, population density is not significantly correlated with broadband penetration. Indeed, one of the world's leading broadband nations, Iceland, has one of the lowest population densities in the world. Furthermore, 5 of the 11 countries ahead of the U.S. in the OECD broadband rankings have lower population densities than the U.S.

While there may be a theoretical reason to think that population density should be correlated with broadband penetration, in real world measurements comparing aggregate performance at the *national* level that is not the case. There is a related phenomenon of "economies of density" that we also examined. In theory, it should be less costly on a per-line basis to deploy broadband to an area that is highly populated than one that is sparsely populated — all other things being equal. But population density is not the relevant metric to capture this phenomenon — as people tend to cluster in cities, regardless of the overall geographical area of a particular country. The relevant metric is "urbanicity," or the percentage of a nation's population living in urban areas or clusters. We found the US has a relative high rate of "urbanicity"—very similar to that of South Korea. This is why on the whole—despite our vast, rural areas—our population density doesn't have a strong influence on national broadband performance.

When the relationship between urbanicity and broadband penetration is examined, there's only a very weak, statistically insignificant correlation. Countries like the Netherlands and Switzerland have lower percentages of their population living in urban areas than the United States yet have higher broadband penetration rates. Similarly, countries like New Zealand and Germany have higher percentages of urban population than the United States but lower broadband penetration levels. In total, 6 of 11 countries ahead of the U.S. in the OECD broadband rankings have lower percentages of their population living in urban areas.

In short, geographic factors alone cannot explain why the United States lags behind. Factors like average household income, income distribution, public policy, and market competition play a far bigger role.

That is not to say that there we have the perfect policies to specifically address the marketplace problems of rural America. It is true that a nation to nation comparison does not show population density to be a significant factor in evaluating overall performance. However, our rural populations do have service characteristics that are unique in distance and density that make ubiquitous Internet access challenging domestically compared to urban areas. The single most important policy change we can make is to follow up on the Joint Board's recommendation to transition the USF programs to cover broadband. These programs—while in need of reform—should do for broadband in rural areas what they have done for telephone service. Additionally, we should look to new wireless technologies (such as utilizing the broadcast "white spaces") to open new avenues for service delivery that are less costly to build out. These wireless services will not be a substitute for wireline infrastructure, but they may prove quicker to deploy and bring the benefit of mobility and a carrier of last resort to unserved areas.

3. **The Internet Tax Freedom Act, which is set to expire on November 1, 2007 places a moratorium on the ability of state and local governments to impose: (1) new taxes on Internet access; or (2) any multiple or discriminatory taxes on electronic commerce. I am greatly concerned that allowing the moratorium to expire would be devastating to small businesses. For example, increasing the cost of Internet access to home-based consumers would reduce consumer demand and reduce telecommunication companies' incentive to build rural broadband networks. Without broadband, rural small businesses would be unable to access high-speed Internet and truly compete in the global marketplace. What would be the impact on small businesses if we begin to allow state and local governments to tax Internet access, please describe? From your perspective is there merit to my concerns?**

The concerns about taxing Internet access are certainly merited. The market for Internet access is not yet mature, and there is good reason to believe that (at least in the short term) it is relatively price sensitive. Increases in price because of taxation could result in declining demand, or worse, a decline in subscribership. However, this question should be revisited periodically, as price elasticity will change and the needs and reasons for moderate state and local taxation will evolve and merit attention in their own right. The Congress has now acted on this question and reinstated a temporary moratorium on Internet taxation. It should be noted that the potential harm to small businesses from taxation on Internet access is nowhere close to the level of damage caused by a lack of competition in the broadband market and the resultant high prices and paucity of suitable, affordable services.

4. **What makes the Internet special is that it costs virtually nothing to create a website and send and receive electronic files and other data. Small businesses can sell their products in the same manner as their large corporate competitors, without the inhibition of geography – something that's critical to businesses in rural locations. Cost is not a barrier to access, and therefore those who use the Internet can access virtually any content from an infinite universe of on-line sources, commercial or otherwise. What concerns would you have with an Internet where cost is a barrier to access? Do you feel we are headed in that direction? Why or why not?**

It is imperative that discriminatory costs do not become a barrier to entry for small business driven commerce on the Internet. And there is most certainly concern that we are headed in that direction. This goes to the threshold issue of Network Neutrality that the Congress has been debating for two years. The FCC and the courts have opened the door for the network owners to alter the architecture and the cost structure of the Internet. The network owners have announced their intention to do so to Wall Street, complete with plans to charge discriminatory prices for quality of service. This "tiering of the Internet", if it is permitted, would be a disaster for small businesses and the free wheeling marketplace for online commerce. It would shut down what is best about the Internet as we know it. It is critical that Congress act to reinstate Network Neutrality.

In the words of Internet architect Vint Cerf, the Internet allows "innovation without permission." This genius of the network has proven to be a wonderland for entrepreneurs, many of them small businesses. It is critical to remember that the Internet's name brands of today were just "good ideas in garages" a decade ago. College kids created Google. A hobbyist conceived the idea for eBay. A teenager wrote the code for Instant Messaging. Some of the most popular sites on the Internet right now— MySpace, FaceBook, and YouTube — didn't exist three years ago. This technological revolution keeps turning because the Internet is an unrestricted free marketplace of ideas where innovators rise and fall on their merits.

The laws that protect this free market are Network Neutrality rules. Without the rules, innovators are at the mercy of the network owners saying who can and cannot succeed. Think about the repercussions of simply raising money from investors in a world without Network Neutrality. How many venture capitalists will embrace a business plan if the first line reads: "Strike a favorable deal with AT&T"? It is simply a non-starter for entrepreneurs that will stifle innovation.

This nightmare scenario is hardly hypothetical. Hardware manufacturers currently advertise routers that have the ability to investigate the packets flowing onto a network to determine the origin of the content or application. Comcast was recently revealed to be using a technology that blocks particular applications that might one day compete against cable television. Complaints are pending at the FCC on this matter. The danger of content discrimination is clear. If the content comes from a "preferred" provider that has made a deal with the network, it is guaranteed quality of service. If the content is from an unaffiliated source, the router can de-prioritize the content and degrade the service. Network operators are already planning to manage bandwidth to maximize revenue streams through discriminatory deals with third-party providers. This distorts the market, undermines competition, and smothers innovation. It must not be permitted.

5. **Rural America, which is home to nearly 25% of the nation's population, comprises 75% of this nation's land mass. However, large parts of rural America are losing out on jobs, economic development, and civic participation because of inadequate access to high-speed Internet. The Federal government has initiated programs like the Rural Development Broadband Loan Guarantee Program, but from a practical standpoint, these efforts appear to be wasted. What are some specific things the Federal government can do to encourage rural broadband deployment? Are improved broadband data gathering procedures and Universal Service Reform two ways that we can improve our broadband penetration rate in rural America? Do**

you believe a wireless connection can be an effective substitute for DSL or cable wireline in rural communities?

The persistent gap between rural and urban broadband penetration has a number of causes. Most of them come down to a simple value proposition. Is the broadband product that is available worth buying (assuming there is one to buy)? The core policy problem is to find ways to make that product more valuable. At the most basic level, if broadband is totally unavailable, its value is zero and our task is simple---we must find ways to get broadband to areas that have no service whatsoever. However, these areas are shrinking and the absence of service does not constitute our most challenging difficulty. The second level problem is that the available broadband service is too expensive and/or too slow---that is, it is not valuable to consumers and so they do not subscribe in large numbers. This is particularly true in rural areas that are also low-income areas. Here, we will find that computer ownership is also low, as is technology training, a situation that does not lend itself to spending limited disposable income on broadband. These are some of the reasons why we have a low broadband penetration rate, despite service availability to over 90% of households.

The answer is to bring more and better service to these areas coupled with social programs that make broadband more valuable to consumers. The first order of business is certainly better data collection. We cannot manage what we do not measure. Currently, we are probably wasting a significant amount of USF money targeted at areas that are less needful than others. But we do not know, because we do not have data about subscriber numbers that are more detailed than the state level. We need granular data that we can cross-reference with demographic information from the Census Bureau. We need to know which ISPs are performing well compared to others and figure out why. We need to know where there is competition, where prices are high and low, and where speeds are adequate and where they lag. And we need to keep collecting this information longitudinally in order to measure our progress.

Armed with this information, we can confidently transition the USF programs from dial tone to broadband. It is a long awaited change and the Joint Board has begun to move in this direction. USF broadband programs will be the most important factor in increasing rural broadband penetration. But they must be paired with programs that bring computers and training to rural households. And they must be paired with policies that encourage competition in markets where it is economically feasible. If market forces can deliver faster, cheaper broadband and improve the value of the products for consumers, we should tap those forces for all they are worth and supplement them with USF support.

While it is possible that wireless broadband may one day be a substitutable product for DSL and cable modem, it is not a substitute today. Further, there is little reason to believe that it will become a substitute in the near future. Wireless broadband is typically twice as expensive and half as fast (at best) as a wireline broadband product. It has the advantage of mobility, but consumers use wireless broadband for different purposes than home broadband connections. According to the FCC, almost 90% of wireless broadband products are purchased by business consumers. This clearly indicates the state of the market today. Judging the market as it stands today, it would be very risky to bet that wireless will catch up to wireline and be a substitutable product that eliminates the need for wireline infrastructure. On the contrary, high capacity wired networks appear to be the clearest path back to global competitiveness for US broadband markets. Wireless will add a product market for mobile devices.

6. **One thing that I find exciting about the Connect Kentucky project discussed during the Committee hearing is the tremendous amount of data that will begin to allow us to test the effects of different policies. However, in your testimony you advocate a number of policy changes by citing that the data we do have proves that there is an immediate problem that must be addressed. In your opinion is this prudent public policy? Shouldn't we first improve data collection efforts and evaluate whether proposed policies are likely to address the above mentioned problems?**

There is indeed a great need for more and better data to assess the problems in the broadband market and craft targeted solutions for micro-level problems. There will be a variety of policies that we should not enact until we have more information. However, the existing data show us quite clearly where the macro-level problems are and point to immediate actions we can take to begin addressing them. The data gaps are at the local and state level with regard to deployment, speed, price, and penetration rates. We have very useful data at the national level, particularly on subscriber numbers as well as reliable figures for price and speed. We can break out the two types of data/solution pairings as follows:

At a macro-level, the existing data is sufficient to demonstrate the absence of meaningful competition in the marketplace for broadband access. The FCC cannot tell us much about the specifics of broadband adoption in different zipcodes, but the numbers are accurate when it comes to showing (at a national level) which technologies account for most of American broadband connections. We can see that well over 90% of connections are controlled by the regional duopolies of cable and telephone companies. Wireless connections, while increasing, are substantially business connections, and they remain expensive and slow, putting them outside true competition with wireline products. We can also use various international and commercial data sources to compare prices, speeds, and penetration rates to show that the US is considerably behind its global competitors in each of those metrics. While we may not be able to precisely identify just how far behind we are, the trend lines are clear. These data points all suggest that policies that promote competition and deployment are immediately in order. Conversely, any current policy decision that bears the risk of *reducing* competition should be halted.

This leads to recommendations for pro-competition policies such as opening the broadcast "white spaces", protecting the rights of municipalities to offer broadband, and conditioning spectrum auctions on pro-competitive conditions. Further, these data points suggest against any reforms of special access regulations that may result in the elimination of competitors. Finally, the policy rationale for Network Neutrality is clear and present—a small group of network owners have market power over access to the Internet.

At a micro-level, there is a strong need for new data to craft local solutions. For example, the USF programs should be transitioned to broadband. However, targeting the funding to the areas that need it most will be difficult without more granular information about what is happening in local markets. Similarly, efforts to target other subsidies (such as tax incentives) should be reserved for systematic use when we can direct the resources appropriately. Finally, the need for social programs that provide computers and training to those areas with broadband access but low broadband penetration will be best understood with better data. In all these cases, however, the data is sufficient to suggest which direction we should be moving toward, even if it does not permit specific decisions. Notably, we cannot wait until a new batch of better data arrives to begin the long, slow process of USF reform.

7. **Can wireless Internet be the solution to the lack of deployment in rural areas? Why or why not?**

As I have noted in my previous answers, the state of wireless broadband today does not suggest it is a viable solution for rural broadband deployment in the near term as a substitute for wireline connections. Wireless has many advantages, and it should be deployed in rural areas, just as it is in urban areas where wireline connections are more prevalent. Mobile communication through wireless broadband promises to become more and more essential to consumers. However, the price per unit of speed in wireline connections appears likely to outstrip the capabilities of wireless for the foreseeable future. If, as many believe, we are headed toward a fiber-optic future, there will need to be considerable advances in wireless in order to bring it into parity with a wired infrastructure. That said, if wireless is the only technology that is cost-effective for deployment in some rural areas (and the alternative is nothing), then it should certainly be promoted.

But the bottom line is that wireless and wireline broadband are developing into two distinct product markets. Each has its advantages and disadvantages. Each is valuable to consumers for different reasons. Each carries social and economic benefits that we all want for our communities. Consequently, we should be thinking about a broadband policy that brings both of these technologies to rural areas, just as we are in urban areas.

RESPONSE TO POST-HEARING QUESTIONS FROM THE U.S. SENATE COMMITTEE ON SMALL BUSINESS AND ENTREPRENEUERSHIP BRIAN MEFFORD

Brian Mefford
President & CEO
Connected Nation, Inc.
bmefford@connectednation.com

Response to Post-Hearing Questions from the U.S. Senate Committee on Small Business
and Entrepreneurship

October 31, 2007

1. **One argument made by the FCC and others is that we don't need any
   consumer safeguards for Internet users – including "net neutrality" – because
   there is robust competition in the broadband market. However, most people
   in any state are lucky if they have more than one choice for broadband
   provider. Do you consider the broadband marketplace to be highly
   competitive? If the market place is competitive, why do we continue to
   experience a lack of broadband penetration, specifically in rural areas? I hear
   a lot of talk about how wireless broadband from a cell phone tower is a viable
   alternative for DSL or cable modem. Do you believe a wireless connection is a
   capable substitute for DSL or cable Internet?**

   The marketplace is competitive in markets that can sustain competition. Every
   American needs broadband availability, and a large part of Connected Nation's
   mission is to ensure that every resident and business has some form of broadband;
   however, if we force additional providers in rural areas where there is not a market
   for them, the businesses would fail. Competition is certainly ideal, and that is the
   reason Connected Nation's work is focused as much on driving adoption of
   broadband and demand for services as it is on mapping the broadband gaps. The
   better consumers can effectively demonstrate demand to service providers, the
   more likely it is that providers will see the business case for offering additional
   service. To this end, oftentimes after Connected Nation convinces a provider to
   deploy service in an unserved area, competition quickly follows as a result of the
   immediate response from consumers to subscribe to broadband at higher-than-
   expected rates for a rural area. This high adoption rate is a result of the "demand
   side" local technology planning process and awareness building that Connected
   Nation facilitates and empowers in every community across a state.

   During ConnectKentucky's initial program development stage, the resounding
   message from both large and small providers was, "We want to help serve the
   unserved areas, but until we get those who are already served to actually subscribe,
   we cannot continue to invest in areas where we will lose money." When we
   conducted surveys to understand the barriers to Internet and broadband adoption,
   the results showed that the top reasons people did not subscribe were not associated
   with cost of the service or lack of availability, but rather that people did not own a
   computer or did not understand why they needed broadband. It was research such
   as this that laid the groundwork for the development of a demand-driven model for
   broadband expansion.

The good news with competition is that our research shows that in areas where there is little to no competition, consumers are not seeing a difference in price or satisfaction of broadband service. That is, consumers pay around the same amount for broadband service regardless of the number of providers, and consumers report high satisfaction levels across the board.

Recent research released by the Information Technology & Innovation Foundation supports Connected Nation's research, indicating that lack of competition in broadband markets has not produced the negative effects for consumers that might be expected.
http://www.itif.org/files/BroadbandCompetition.pdf

In response to the question of whether wireless broadband is a viable solution, there is an important distinction to make between fixed wireless broadband and mobile wireless broadband. Fixed wireless is a reliable and affordable form of broadband that can produce speeds at or near wireline broadband at a fraction of the capital investment required for running cables or upgrading telephone facilities. Fixed wireless has been deployed in many rural communities where there is no other economically sustainable solution, and these rural residents and businesses are now using the Internet in life-changing ways – through e-government, teleworking, online businesses, workforce development, online education, and the list goes on. However, mobile wireless broadband is a very different form of service. Mobile broadband is accessed only through mobile devices, and there are limitations to this service – particularly in rural areas. While this service (in the areas it exists – which are mainly metropolitan areas) provides another advanced service option for consumers, and is certainly a great product, Connected Nation does not consider this technology, as it is available today, to be a "replacement option" for primary home broadband use, and this service is not included in the broadband availability numbers reported by Connected Nation state programs.

2. **The U.S. is made up a unique geography and population density when compared to other countries. With 32 people per square kilometer the U.S. ranks 12th in population density and 15th in terms of broadband penetration. However, other countries who are leading the way in broadband penetration have much higher population densities. For example, Korea is ranked 2nd in broadband deployment and has a population density of 481 people per square kilometer. Is population density a key factor in broadband deployment, and if so, are we implementing proper policies to suit our unique geography? Why or why not?**

Population density is certainly a key factor in broadband deployment; however, it is not the only one. Terrain is another primary factor. Certain technologies work better in particular topographies. The critical question that Connected Nation must ask when working with communities and providers to serve unserved areas is this: "What technologies are available that can provide the most cost effective *and* *sustainable* solution for this community and its needs?" The geography of each state (and often each county within a state) is different, and it doesn't make sense to implement a "one solution fits all" approach for a country like the United States. The ConnectKentucky program has proven that statewide solutions can be effectively achieved if communities are provided with resources for local technology planning and if providers can unite behind a common purpose for filling the broadband gaps in a way that makes economic sense. It is these two pieces coming together that drives local solutions to local problems – across an entire state, and ultimately across all states.

Senate Bills 1190 and 1492 provide a policy framework and resources for state based programs to effectively implement accurate broadband mapping in coordination with local technology planning. These bills establish a simultaneous supply and demand process that drives broadband deployment and also improves broadband use.

3. **The Internet Tax Freedom Act, which is set to expire on November 1, 2007 places a moratorium on the ability of state and local governments to impose: (1) new taxes on Internet access: or (2) any multiple or discriminatory taxes on electronic commerce. I am greatly concerned that allowing the moratorium to expire would be devastating to small businesses. For example, increasing the cost of Internet access to home-based consumers would reduce consumer demand of and reduce telecommunication companies' incentive to build rural broadband networks. Without broadband, rural small business would be unable to access high-speed Internet and truly compete in the global marketplace. What would be the impact on small businesses if we begin to allow state and local governments to tax Internet access, please describe? From your perspective is there any merit to my concerns?**

While Connected Nation has taken no official stance on the Internet tax moratorium, it does recognize the inhibiting factor that increased taxation can have on the broadband market and its development. At the same time, however, because Connected Nation's public-private partnership model depends on state and local government support, Connected Nation also understands the desire of state and local governments to increase their tax revenue. Due to the 4-year temporary moratorium that became law this week, Connected Nation will continue to ascertain what it believes to be right balance between the necessary freedom of the market to make investments in infrastructure and the needs of government.

4. **Rural America, which is home to nearly 25% of the nation's population, comprises 75% of this nation's land mass. However, large parts of rural America are losing out on jobs, economic development, and civic participation because of inadequate access to high-speed Internet. The Federal government had initiated programs like the Rural Development Broadband and Loan Guarantee Program, but from a practical standpoint, these efforts appear to be wasted. What are some specific things the Federal government can do to encourage rural broadband deployment? Are improved broadband data gathering procedures and Universal Service Reform two ways that we can improve our broadband penetration rate in rural America? Do you believe a**

<u>wireless connection can be an effective substitute for DSL or cable wireline in rural communities?</u>

Passage of S. 1190, the Connect the Nation Act, or S. 1492, the Broadband Data Improvement Act, is the first action the federal government should take to encourage national broadband deployment. Both pieces of legislation, by assisting states with state-based programs that include a public-private partnership such as ConnectKentucky, would improve both broadband data and deployment.

As outlined above, fixed wireless broadband is a cost-effective and sustainable solution for many rural communities, and can be an effective substitute for DSL or cable.

5. **The benefit of broadband deployment is undeniable and ConnectKentucky's efforts at the municipal level through public-private collaboration should certainly be commended. As a matter of fact, the Government Accountability Office (GAO) has singled out ConnectKentucky's broadband data gathering efforts, noting that they are a significant improvement over the Federal Communications Commission's zip code data. As I understand it, the FCC currently relies on zip codes to determine if broadband is reaching all Americans. If one subscriber in a zip code receives broadband, the FCC assumes that broadband is available throughout the area. But I have often heard an analogy that this is like saying, if one person in a zip code drives a Mercedes, we can assume throughout the zip code, everyone drives a Mercedes. <u>How does ConnectKentucky, and now Connected Nation's, mapping methodology improve upon the FCC's current data gathering efforts? While ConnectKentucky has shown great strides in broadband deployment, what challenges were you faced with throughout the initial start-up and how have you overcome them? Are you still facing challenges today? If this system is truly expandable to all states, would they too, encounter similar obstacles?</u>**

The Connected Nation mapping methodology is rooted in a community driven technology planning process that creates demand for broadband and IT services, which in turn drives the supply. The success of this model is partially dependent upon its comprehensive nature – it systematically addresses both supply and demand through specific programs and processes that include not only statewide mapping of the broadband "gaps" and the creation of market intelligence maps for providers for unserved areas, but also targeted application development and awareness building which is enabled by county level research on business and residential use of technology and barriers to adoption, computer distribution programs for underprivileged children, and the grassroots creation of a local strategic technology plan for every county. All these pieces are combined and used in collaboration with providers and communities to leverage resources and find solutions for each unserved area in Kentucky – to ensure that all residents and all businesses have the ability to subscribe to broadband – and to simultaneously bring about dramatic increases in technology use for economic and community development.

Connected Nation developed this model after years of research and discussions with Kentucky providers who told us that they want to help serve the unserved, but they can't invest further until they increase take rates in those areas where they have already invested. This type of information led us to the development of what is now the Connected Nation model which creates specific and measurable programs to drive demand and increase technology literacy.

Broadband mapping is one of the foundational tools Connected Nation uses to implement this process across a state. Connected Nation works directly with broadband providers to understand exactly where broadband service is offered, down to a street level. This process involves relationship building and taking the time to understand how this information can be provided in the most safe and efficient way. It is part of Connected Nation's job to make this process easy and valuable for providers.

The broadband "gap identification" map, however, is simply the first step in the process of finding broadband solutions for unserved areas. Once the gaps are identified, Connected Nation develops a series of maps that drill down into those unserved areas and analyze those markets. This includes analysis of household densities at a very granular level, terrain analysis, identification and evaluation of existing and potential infrastructure that could be used for fixed wireless deployment such as water and cell towers, and mapping of proposed infrastructure such as water lines, sewer projects, and future roads. It is this analysis, used in combination with Connected Nation statistical research on technology adoption among local residents and businesses, which provides the basis for discussions with providers and communities in finding solutions.

Early on in the research and development of the ConnectKentucky model, we quickly learned that many broadband providers are not willing to hand over proprietary and confidential infrastructure data to the state. So it was decided to house the operation within the nonprofit 501(c) 3 that is now called Connected Nation. Although it took a good deal of time and effort on the front end, this process now works well because providers have built a trust with Connected Nation and its ability to hold data confidential as well as produce high quality GIS maps that are useful for providers in encouraging deployment in the hardest to reach areas. Perhaps most importantly, Connected Nation staff understands both the industry as well as community needs. They serve as the experts in the middle who can build partnerships at both the state and local levels to find the best solutions for those "last mile" areas.

Other states such as Tennessee are now working with Connected Nation to build public-private partnerships similar to ConnectKentucky. Fortunately for these states, the early research and development was done in Kentucky, and Connected Nation is now able to help implement more cost-effective solutions in a shorter period of time. Although the environment of each state is different, and thus the details of implementation are different, these states are reaping the immediate benefits of using the resources of an organization that invested years to build provider trust and to develop a model that works for any state.

6. **My home state of Maine has recently adopted the Connect Maine Authority, an ambitious initiative committed to ensuring that 100% of Maine communities have wireless coverage by 2008. Like ConnectKentucky, Connect Maine attempts to create a public-private partnership for expanding broadband deployment. How can we learn from what's going on with ConnectKentucky and how successful can the program be in other states? Should this type of public-private partnership be administered at the state level or can it work on a national level?**

Having seen successful results in our state-based efforts, Connected Nation believes strongly in the critical role of public-private partnerships in improving broadband availability and adoption. Neutral non-profits such as Connected Nation play an indispensable role in bridging the gap between the needs of business and consumers, government and the free-market. Connected Nation's success in bringing together communities and providers is predicated on a model that uses (and effectively protects) proprietary infrastructure data to produce broadband "gap identification" maps at a neighborhood and street level, and makes these maps publicly available to consumers, policymakers, and providers. Just as importantly, the effective use of the maps to fill the broadband gaps is firmly rooted in the "demand side" grassroots technology planning teams in every community across a state.

Connected Nation's extensive research laid the groundwork for the development of a demand-driven model for broadband expansion. By using statewide demand creation and local technology planning in every community, the model benefits both providers and the state. Take-rates in served areas go up, revenue goes up, investment dollars go up…and then providers are vested in the program and are often willing to move outside their comfort zone to help unserved areas. Meanwhile the generation of demand in these unserved areas often creates a business case for investment where before there was none. And increased technology adoption throughout the state increases the workforce development skills of the citizenry, makes businesses more productive, improves healthcare and education, enhances government services, and creates a better way of life.

As such, a key component of the ConnectKentucky program is its demand-driven model whereby statewide demand generation drives supply into unserved areas. Another critical component is the time that was invested to develop relationships with providers to assure them their sensitive broadband infrastructure data would be protected and used in ways that benefit the state and its citizens, but would also benefit them and their counterparts by creating market intelligence maps to fill the broadband gaps. The non-governmental status of ConnectKentucky allows for nondisclosure agreements to legally ensure that provider data are held confidential. Fortunately, these relationships with providers that developed over years and through countless discussions are now being used to enable a similar model in other states through the non-profit Connected Nation, which was formed in response to other states' requests for help with broadband expansion.

Yet another key feature to ConnectKentucky's work continues to be its research-based approach to strategic development, which enables the creation of targeted programs and applications that fit the needs of each specific community. Beyond accurately measuring the inventory of broadband services to increase investment, ConnectKentucky's county-level statistical consumer surveys guide and direct the local planning teams to create applications that are relevant to local citizens and businesses, and it offers insight for providers, policymakers and local leaders into consumers' technology adoption and usage patterns. This rich information on consumer needs enables the development of programs such as No Child Left Offline – a computer distribution program for disenfranchised populations – to target people who are most in need of computers.

While Connected Nation is aware of efforts to use government entities in the role of a non-profit partner in a public-private partnership, based on its positive experiences in its state-based programs, Connected Nation believes that the successful public-private partnership seen in Kentucky would be inhibited without the critical role of a neutral non-profit presence.

Finally, due to the geographic, cultural, and economic differences that exist between the states, and the highly local demand-driven work that is a necessary element of any project that seeks to replicate Connected Nation's success, Connected Nation believes that similar efforts are best administered on a state-based level rather than as one, national project.

7. **In Mr. Scott's testimony, he applauded programs like ConnectKentucky. However, he stated that there are limitations to your model. He stated that the data your program collects is exclusively proprietary. Also, he stated that your program doesn't collect information about price and speed of broadband connections. <u>Do you consider these real concerns? If your program was instituted on a nationwide basis wouldn't this hurt the effort to create a big picture analysis of the entire country?</u>**

While ConnectKentucky and other Connected Nation programs do establish non-disclosure agreements with providers to protect proprietary data such as the exact locations of competitively sensitive infrastructure and equipment, nearly all of the data collected by Connected Nation becomes publicly available when maps are produced. In fact, that is the entire reasoning behind the model – to figure out a way to analyze proprietary and competitively sensitive data in such a way that providers can be comfortable with – in an effort to publicly produce an assessment of where broadband truly exists so that we can work together to fill the gaps and increase technology adoption at the same time.

Further, the mapping dataset is only one piece of the range of data collected by Connected Nation. Just as importantly, we conduct extensive telephone surveys among residents and businesses to understand technology use and barriers to use –

the results of which are public. Connected Nation's qualitative data collection with local technology planning teams is also public, and the strategic plans of those groups are public. All of this information is posted on the websites of Connected Nation's state programs.

Connected Nation does collect extensive information about price and speed of broadband. Mr. Scott was likely referring to the fact that we do not collect this information from providers, but rather, from consumers themselves. Collecting this type of data from providers is not only inefficient, but it does not give a clear indication of what residents and businesses are actually spending or using. Connected Nation collects price information directly from consumers through statistical telephone surveys at a county level and through local teams. Speed data is collected through online consumer speed tests, which follow in the footsteps of the Communication Workers of America's Speed Matters campaign. These state-based tools enable granular data collection of *actual* speeds (as opposed to advertised speeds) with a representative sample for all communities.

WALLSTEN RESPONSES TO QUESTIONS FROM SENATOR OLUMPIA SNOWE REGARDING BROADBAND AVAILABILITY

1. One argument made by the FCC and others is that we don't need any consumer
   safeguards for Internet users – including "net neutrality" –because there is robust
   competition in the broadband market. However, most people in any state are
   lucky if they have more than one choice for broadband provider. Do you consider
   the broadband marketplace to be highly competitive? If the market place is
   competitive, why do we continue to experience a lack of broadband penetration,
   specifically in rural areas? I hear a lot of talk about how wireless broadband from
   a cell phone tower is a viable alternative for DSL or cable modem. Do you believe
   a wireless connection is a capable substitute for DSL or cable Internet?

Contrary to conventional wisdom, broadband penetration in the U.S. is not low compared to
other developed countries. Residential penetration in the U.S. was just over 50 percent in early
2007. This household penetration rate compares to an average of about 28 percent for EU
countries. As Figure 1 below shows, relative to EU countries, U.S. residential broadband
penetration is below only the Netherlands and Denmark.[1]

Figure 1
Household Broadband Penetration in the EU and US

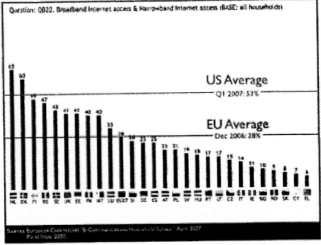

---

* I thank Stephanie Hausladen for excellent research assistance in preparing these responses. However, the opinions
expressed in this document are mine alone and I am solely responsible for any mistakes.
[1] Household penetration in Japan, according to the OECD, is about 65 percent. Beginning in 2004 household
penetration in Korea was reported in a way that made it impossible to compare to other countries. Nevertheless,
household penetration in Korea was about 67 percent in 2003, so it must be higher than in the U.S.

These numbers differ from the standard OECD estimates because the OECD attempts to count both residential and business connections. Unfortunately, it is difficult to count business connections. Most computers in office buildings in the U.S. connect through special access lines. Neither the FCC nor the companies that supply the special access lines know how many computers are connected. Thus, these connections are excluded.

The only reliable way to estimate residential penetration is through surveys. The EU has conducted surveys of about 26,000 households annually for several years. The Pew Internet and American Life Project surveys about 2,200 households in the U.S. each year (several private research firms also conduct similar surveys).[2] The OECD itself now recognizes these issues, and publishes household survey information in addition to its rankings.[3]

Consumers are adopting broadband remarkably quickly. It is important to remember that broadband is relatively new, and is being adopted faster than most other new technologies. Figure 2 shows that American households are adopting broadband faster than they adopted cable television, personal computers, CD players, or VCRs.

Figure 2
Percentage of Households With Selected Technologies[4]

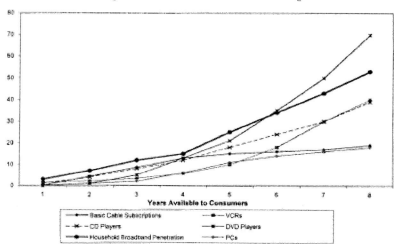

[2] http://www.pewinternet.org/pdfs/PIP_Broadband%202007.pdf
[3] The OECD's new "Broadband Portal" is available here: http://www.oecd.org/sti/ict/broadband.
[4] Sources: Data for PCs, VCRs, DVD players are from the Consumer Electronics Association (http://www.cbrain.org/). Data for basic cable subscriptions are from Cable: A.C. Nielsen Co. as reported by the National Cable & Telecommunications Association (NCTA http://www.ncta.com/ContentView.aspx?contentId=3577 (cable figures begin in 1972, which is widely regarded as the beginning of the cable industry's modern history as a satellite retransmittal service), and household figures are extrapolated from the U.S. Census). Broadband data are from Point Topic.

The market appears to be reasonably competitive and continues to become more so. The FCC reports that DSL is available to 79 percent of all households that have telephone service from an incumbent local exchange carrier and that cable broadband is available to 96 percent of all households that can subscribe to cable television.[5] In addition, the FCC data show that the percentage of zip codes with four or more broadband providers has increased from 31 percent in 2001 to 83 percent in 2006. The FCC's zip code data have some well-known flaws, but they show a clear trend of increasing competition (Figure 3).[6]

Figure 3
Percent of Zip Codes With Four or More High-Speed Providers

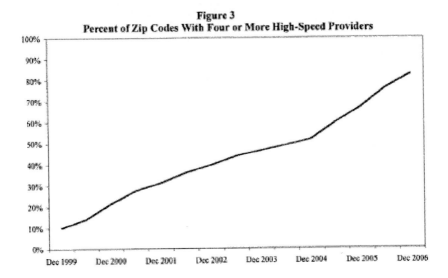

While competition is increasing, few consumers have more than two choices for wired broadband connections (DSL and cable).[7] Additional competition, therefore, is likely to come from wireless providers, at least in the short- to medium-term. Consumers have many choices for wireless broadband. Both Sprint and Verizon Wireless offer wireless broadband, and at least two satellite companies (Wildblue and HughesNet) offer coverage that blankets the country. In

---

[5] FCC. "High Speed Service for Internet Access." October 2007. Table 14.
[6] In particular, the FCC counts a provider as offering service in a zip code no matter how many customers it has. This method of counting overstates the degree of competition.
[7] Some consumers may have additional choices from "overbuilders"—competing cable networks such as RCN—but these serve only a small share of consumers.

addition, a large number of small Wireless Internet Service Providers (WISPs) provide service over relatively limited geographic areas.[8]

Wireless broadband is slower and more expensive than wired connections, but speeds are increasing and prices are falling. For example, in 2005, Verizon's wireless broadband service cost $79.99 per month for a 400-700Kbps connection.[9] Today, Verizon and Sprint offer download speeds of 500Kbps-1.5Mbps for $49.99-$59.99 per month.[10] In two years speeds have approximately doubled and prices have decreased by 25-40 percent. Satellite companies are upgrading their broadband offerings as well. Wildblue offers a maximum speed tier of 1.5Mbps for $79.99 per month.[11]

While wireless service continues to improve, it cannot currently match wireline (especially fiber) speeds. Comcast offers a cable connection of 6Mbps for $57.95 per month and Verizon offers 3Mbps with DSL technology for $29.99.[12] As a result, wired and wireless cannot be perfect substitutes. The question is how substitutable are they?

Many subscribers have little need (currently) for very fast connections.[13] As a result, for most consumers wireless is potentially a good substitute for wired broadband and it thus puts competitive pressure on wired broadband providers. Prices for wireless broadband have already decreased considerably as speeds have increased. As additional spectrum is introduced into the market, technology improves, and competition increases, prices will continue to come down.

---

[8] See http://www.bbwexchange.com/top10wisps.asp or http://www.wispa.org/
[9] http://www.prnewswire.com/cgi-bin/stories.pl?ACCT=104&STORY=/www/story/08-29-2005/0004095512&EDATE=
[10] http://www.sprintbroadband.com/download_speeds.html, http://b2b.vzw.com/broadband/overageurea.html
[11] http://www.wildblue.com/getWildblue/doServiceAvailabilitySearchAction.do
[12] http://www22.verizon.com/content/consumerdsl/plans/all+plans/all+plans.htm, http://www.comcast.com/Shop/Buyflow/default.ashx
[13] The Pew 2007 Home Broadband Adoption Report finds that the most common online activities among household with broadband connections is email (95% of home broadband users), "look[ing] for information about a hobby or interest" (89%), getting news (79%), doing research for one's job (57%), and looking for information on Wikipedia (42%).

2. The U.S. is made up of a unique geography and population density when compared to other countries. With 32 people per square kilometer the U.S. ranks 12th in population density and 15th in terms of broadband penetration. However, other countries who are leading the way in broadband penetration have much higher population densities. For example, Korea is ranked 2nd in broadband deployment and has a population density of 481 people per square kilometer. Is population density a key factor in broadband deployment, and if so, are we implementing proper policies to suit our unique geography? Why or why not?

Population density is important, since it is less costly to provide service when people live closer together. Many studies have shown a statistically significant positive correlation between various measures of population density and broadband penetration. While there is no evidence of an overall market failure, it is certainly true that some rural areas lack coverage.

We do not yet have good evidence on what policies can improve access in rural areas. Existing approaches, such as the Rural Utilities Service, offer no studies demonstrating that their grants have efficiently improved access. In its 2005 annual report (the most recent available), RUS notes that "[s]ince 2001, Rural Development has utilized a variety of loan and loan guarantee programs to provide over $3 billion in funding and assist over 1.3 million rural subscribers in accessing new broadband technologies."[14] These numbers suggest that RUS has spent about $2,300 per person connected, which implies that the program—based on the agency's own numbers—is not especially efficient.

I will discuss the universal service program in answer to question 7 below.

3. The Internet Tax Freedom Act, which is set to expire on November 1, 2007 places a moratorium on the ability of state and local governments to impose: (1) new taxes on Internet access; or (2) any multiple or discriminatory taxes on electronic commerce. I am greatly concerned that allowing the moratorium to expire would be devastating to small businesses. For example, increasing the cost of Internet access to home-based consumers would reduce consumer demand and reduce telecommunication companies' incentive to build rural broadband networks. Without broadband, rural small businesses would be unable to access high-speed Internet and truly compete in the global marketplace. What would be the impact on small businesses if we begin to allow state and local governments to tax Internet access, please describe? From your perspective is there merit to my concerns?

I agree with your concerns. I see two problems from an economics point of view. First, to the extent that consumers remain price sensitive when choosing whether to subscribe to broadband services, such a tax would be inefficient and therefore costly to the economy. Second, taxing the service would increase prices and thus reduce demand. To the extent that policymakers want to increase penetration, such taxes would undermine those efforts.

---

[14] The report does not provide any details on how the number 1.3 million was determined, or whether any empirical testing was done to determine whether the program itself was responsible for making broadband available to those 1.3 million people.

4. What makes the Internet special is that it costs virtually nothing to create a website and send and receive electronic files and other data. Small businesses can sell their products in the same manner as their large corporate competitors, without the inhibition of geography – something that's critical to businesses in rural locations. Cost is not a barrier to access, and therefore those who use the Internet can access virtually any content from an infinite universe of on-line

sources, commercial or otherwise. What concerns would you have with an Internet where cost is a barrier to access? Do you feel we are headed in that direction? Why or why not?

The Internet has had a positive effect on entrepreneurship and given small businesses access to new markets. The key question in the net neutrality debate is how to balance incentives to invest in the edges (content) versus in the network. Network neutrality proponents believe all innovation should occur on the edges, while opponents focus on investment in infrastructure. The truth is that in this type of market the right pricing structure simply is not obvious. We should encourage innovation in both content and infrastructure.[15]

We do not yet know what the most successful business models will be, and we do not know what businesses might develop under other pricing schemes. For example, it is possible that a tiered Internet service of the type net neutrality proponents generally oppose would yield new businesses that cannot work without guarantees of very high levels of reliability and speed. Consider letter and package delivery. FedEx succeeded precisely because it is able to offer faster and more reliable service for those who need it.

The vast majority of small businesses impose few, if any, additional costs on the network. They are unlikely to ever pay for more than their own Internet connection. Other businesses might impose very high costs on the network. Peer-to-peer file sharing, for example, can lead to network congestion. Asking those companies to pay for the costs they impose on all other users would increase efficiency.

---

[15] To date (to my knowledge), the only instance of contractual fees between a content provider and a broadband provider is actually one in which the content provider demands fees from the broadband provider rather than the other way around. Broadband subscribers can navigate to ESPN360.com only if their ISP (e.g., Comcast or Verizon) pays ESPN a per-subscriber fee. If a subscriber of an ISP that does not pay the fee goes to ESPN360.com she will see a message that reads, "Your Internet service provider doesn't carry ESPN360.com, so you are missing out on over 2,500 free live events." This reverse violation of net neutrality highlights several points. First, it is not obvious who has market power. In this case, ESPN can charge ISPs, not the other way around. Second, some products may be developed only if the innovator is free to try new pricing models. ESPN360 simply may not exist if it were not allowed to charge ISPs. Third, some ISPs use availability of ESPN360 as a way to differentiate themselves from other ISPs. RCN, for example, promotes the availability of ESPN360 as a reason for consumers to sign up (http://investor.rcn.com/ReleaseDetail.cfm?ReleaseID=269130).

5. Rural America, which is home to nearly 25% of the nation's population, comprises 75% of this nation's land mass. However, large parts of rural America are losing out on jobs, economic development, and civic participation because of inadequate access to high-speed Internet. The Federal government has initiated programs like the Rural Development Broadband and Loan Guarantee Program, but from a practical standpoint, these efforts appear to be wasted. What are some specific things the Federal government can do to encourage rural broadband deployment? Are improved broadband data gathering procedures and Universal Service Reform two ways that we can improve our broadband penetration rate in rural America? Do you believe a wireless connection can be an effective substitute for DSL or cable wireline in rural communities?

[My responses to questions 2 and 7 cover many of these issues].

I do not believe that any empirical evidence demonstrates that rural America has been harmed by inadequate Internet access. To the extent that there are market failures in rural areas, policies should be carefully targeted to address specific problems.

6. In your testimony you cite one problem that Congress could address right now, opening more wireless space for broadband. Many people argue that wireless Internet is both too expensive and too slow. Why do you believe this is one of the more important issues for Congress to address now?

Additional spectrum will make it easier for providers to offer faster broadband services, and additional competition will help drive down prices.

In my testimony, I was referring specifically to the spectrum sold in the Advanced Wireless Services (AWS) auction in August and September of 2006. New entrants are poised to offer nationwide wireless broadband. In particular, with its new spectrum T-Mobile can roll out its 3G network, and a coalition of cable providers also bought spectrum with the intention of offering high-speed networks.[16]

This spectrum was made available by moving government agencies to different spectrum bands. Revenues from the auction were intended to pay for this move. Unfortunately, while these firms have paid for the spectrum, the agencies are not relocating. Their intransigence means that these new entrants are facing delays in offering their new services.

If improving broadband and wireless competition is a goal, then Congress could push the relevant agencies to relocate their wireless use. That action would require no additional investment or policymaking and would yield immediate benefits to consumers.

---

[16] In addition, regional wireless firms MetroPCS and Leap Wireless bought spectrum that will allow them to become national firms.

7. As you well know, in addition to being an economic driver, broadband also enables revolutionary cultural, political, and educational exchange. Many of the pending Universal Service Fund (USF) reform proposals in the House and Senate include provisions that expand the universal service distribution to cover broadband expenses. For example, the "Universal Service for All Americans Act," which I have cosponsored, would create a $500 million fund within the USF to help support deployment in areas currently unserved by broadband service. I believe that the USF should be adapted to help support some level of broadband expenses, but I fear that if done improperly, the USF could skyrocket to a size that consumers and service providers that pay into the fund will be unwilling to bear. Please discuss your opinion on the legislation that has been introduced in the Senate. In your esteemed opinion, what would be an appropriate way to target support for broadband deployment in a way that does not cause an undue expansion of the fund? What benefits would broadband deployment have on the software and hardware industries, small business and other areas of commerce? What would be the socioeconomic benefits of broadband deployment to under served areas?

Universal service high cost fund expenditures are increasing rapidly (Figure 2). The universal service program is flawed. Because it taxes a price-sensitive service (telephone usage) to subsidize a service that is less price-sensitive (access) it is inefficient. By taxing poor urban people and subsidizing wealthy rural people it is inequitable. It has also been largely ineffective. Expanding this program to include broadband would be unwise.

Figure 2

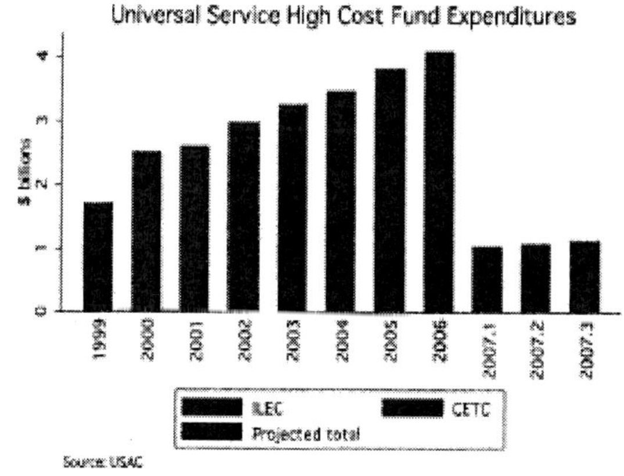

Universal Service High Cost Fund Expenditures

Source: USAC

Cross-subsidies in telecommunications are inefficient and costly to society in large part because they tax usage—such as long distance and mobile—which have relatively high price elasticities of demand in order to subsidize access, which has a low price-elasticity of demand. In other words, our system of funding universal service taxes services for which people are highly price sensitive, causing them to change their behavior and use those services less than they otherwise would. Hausman (1999) estimated that each dollar raised in taxes on wireless services cost the economy between $0.72 and $1.12. Ellig (2005) estimated that taxes on wireless services and interstate long-distance to support universal service reduced economic welfare in 2002 alone—when subsidies were lower than they are now—by nearly $2 billion.[17]

Those taxes, meanwhile, are used to subsidize *access*, which is far less price-sensitive than usage. Rosston and Wimmer (2000), for example, estimated in a detailed empirical analysis that eliminating the high-cost fund would reduce telephone penetration by only one-half of one percent.[18] That estimate is likely to be even smaller today given increased competition and lower costs.[19] In addition, Caves and Eisenach (2007) found no evidence that payments to CETCs increased investment in their service areas.[20] In other words, the evidence suggests that the universal service program is not responsible for increasing service in rural areas.

Given the inefficiency, inequity, and ineffectiveness of the universal service program, expanding it to include broadband would probably be unwise.

---

[17]Ellig, Jerry. 2005. "Costs and Consequences of Federal Telecommunications Regulations" *Federal Communications Law Journal* Vol 58, No. 1pp. 37-102.
[18]Rosston, Gregory L. and Bradley S. Wimmer. 2000. "The 'State' of Universal Service." SIEPR Policy Paper No. 99-18.
[19]Rosston and Wimmer (2000) also point out the inequity of the universal service program: they find that 80 percent of poor households pay into the fund through taxes on telecommunications services they use and get nothing back.
[20]Caves, Kevin W. and Jeffrey A. Eisenach. 2007. "The Effects of Providing Universal Service Subsidies to Wireless Carriers." Criterion Economics. June 13.

SAVE THE INTERNET.com

ACT NOW   the blog   what is net neutrality?   spread the word   about the coalition   press room   donate

Questions from Sen. Kerry: How Can We Connect America?

In a guest blog post, Sen. John Kerry asks for your ideas to repair America's broadband failures and deliver a fast, open and affordable Internet for everyone. Respond to the Senator by commenting in the thread below. Senator Kerry will circle back to address some of your comments and report on developments in Washington.

By Sen. John Kerry

If you talk to anyone in Washington, there's no disagreement that high-speed Internet access is critical to our economic competitiveness, and that a robust and competitive broadband market is key to an affordable and readily available Internet.

For small business, it is critical for the growth of their businesses and the creation of jobs. But everyone agreeing that something's important doesn't get the job rolling in Washington, and there's been precious little actual progress toward improving broadband penetration recently.

Guest Blog Post by Sen. John Kerry

Join the debate

That's why on Wednesday I will chair a hearing to explore the impact of a lack of adequate broadband access on our nation's small businesses. We'll be looking for ways to move closer to making broadband accessible and affordable for every American and every business. We'll hear from advocates for greater broadband penetration (including Free Press' own Ben Scott), and 2 FCC commissioners will be there, as well. You can watch the hearing live at http://sbc.senate.gov.

Speak Out About Universal Access

And, in addition to watching, please put in comments below your own thoughts, recommendations, and plans to improve broadband penetration. You spoke loudly about the need for new competition as the FCC considered spectrum policy. And you had great success.

So let's keep your contributions flowing as we try to get a better Internet in this country. What are your ideas for helping the small businesses and all Americans get faster Internet access?

As activists on this issue, I know you don't have to hear the statistics: more than 60 percent of the country does not subscribe to broadband service — many because they don't have access to broadband Internet service or simply can't afford it. Even in my home state of Massachusetts, a nationwide leader in technological innovation, broadband still has only reached about 46 percent of the public — and that's the fourth-best rate in the country!

It's almost hard to wrap your head around the fact that 7 years into this century, more Americans than not have either no Internet access at all or are still stuck on dial-up. It seems like so long ago that the buzzword was the "information super-highway," but much of America is still bouncing down a country lane. That is just unacceptable.

Restoring America as a Broadband Leader

America's Internet speeds lag badly behind universal standards. The birthplace of the personal computer and the Internet now has far worse broadband penetration than Europe and Japan. Without national broadband access, we're throwing sand in the gears of our economy, placing a technological ceiling of job growth, innovation and economic production.

Small businesses — the backbone of our economy — won't be able to fairly compete. The problem is especially bad in rural areas, and those are some of the areas most in need of economic development in this country.

Some experts estimate that universal broadband would add $500 billion to our economy and create 1.2 million jobs. We need to make universal deployment a national priority to keep America hooked into the increasingly fast global economy, but we can't get that deployment without competition in the broadband market.

Let's Start the Conversation Here and Now

We need a national broadband strategy with a strong federal regulatory framework to encourage competition; companies won't get there on their own. Competition spurs innovation, enhances service and reduces prices. And while we're at it, we need to make efficient and widely available use of the spectrum, a valuable public asset. Much of our spectrum is underutilized, shelved and hoarded by selfish incumbents. Revisions to our spectrum policy must break open the locked portions of our spectrum to maximize that national resource. From drafting "white spaces" legislation to supporting fair spectrum policy, I've advanced and supported a list of measures designed to correct these market failures and increase broadband access.

It's way past time for the country to get serious about this. President Bush has promised national broadband by 2007, and we are inexcusably, tremendously, scandalously short of that goal. Previous generations put a toaster in every home and a car in every driveway as signs of economic progress. To stay competitive, we should strive to do the same with nationwide broadband. Our economy, our businesses and our families are counting on us to deliver.

So, remember to put your recommendations below, and I'll try to circle back after the hearing with another post about what I learned at the hearing and from all of you.

= = = = =
SEP 27 UPDATE FROM SEN. KERRY
= = = = =

Dear all:

I've been reading through your comments, and I just wanted to say that there's a tremendous amount of useful information and suggestions here. I told the FCC yesterday at our broadband hearing that you had some great ideas. Too much to respond to right away, so I'm going to sit down, read them all carefully and get back to you with a full post responding to your great ideas and next steps. Thanks for participating; I learned a lot already from your comments. I'll be back soon.

Sen. John Kerry

This entry was posted on Tuesday, September 25th, 2007 at 1:33 pm and is filed under Uncategorized. You can follow any responses to this entry through the RSS 2.0 feed. You can leave a response, or trackback from your own site.

120 Responses to "Questions from Sen. Kerry: How Can We Connect America?"

1. **auzang Says:**
   September 25th, 2007 at 2:21 pm

   Hi Sen. Kerry-

As a resident of Boston, MA it is unbelievable that in this "world class" city we cannot get DSL, or FIOS into our neighborhood.

The only viable choice for broadband is through cable TV. In order to get the best price for broadband internet I would have to sign up for cable TV. Satellite internet is another option, but the cost is very high.

I still use dial-up, because I don't really have the need for cable TV right now, and to be honest, the cost would be about 3 times higher JUST to get broadband. What's the benefit to that?

Why can't we get DSL and FIOS into more neighborhoods, and why can't there be lower cost, satellite internet options for the "masses", like there is for satellite radio?

One of my biggest complaints is the lack of competition. Why do our lawmakers continue to allow this to happen? We desperately need more competition, more choices, better service, lower prices.

Why don't we have the same choices for internet service providers that people in England do?

Thank you for reading!

2. Zy Says:
    September 25th, 2007 at 4:16 pm

Hey, Senator Kerry,

The internet was born during the time of Local Loop Unbundling, which is what made the internet a success, and which is the regulation that is currently in effect in most of Europe and in parts of Asia where broadband choices are many, speed is great, prices are lower, and real competition exists. People in those countries have DOZENS of choices of broadband providers. And the broadband providers are rolling out their own fiber with no monetary help from those governments.

What a contrast!

Here in the US the Telecomms were given hundreds of billions of dollars in tax incentives to build out broadband by 2006, reneged on the deal, kept the money, and then whined to get price protections in place that effectively KILLED LLU and put thousands of ISPs out of business, using the "we don't have enough money to build infrastructure" argument. Meanwhile they're spending that dough to lobby you folks so they can scr*** us worse. As you can understand, this situation isn't sitting well with us taxpayers.

Now they want to turn the internet into a push media and tell us what we're allowed to access at high speed. They want to be gatekeepers in many ways, and charge for every step of our trip along the network. This will effectively KILL the internet, and our economy with it. It will certainly put this small business web designer out of business.

For more details on all this, do some reading here:
http://mutmachine.com/lotos/

and here:
http://www.nomuswatchdog.org/index.cfm?fuseaction=Ask_this.view&aqkthisid=186

and please interface with Senator Durbin on this. He did a fabulous thing over at OpenLeft, opening the conversations with the techies who have the experience running the network and who easily debunked the FUD coming out of the Telecomm Lobbyists. Oddly enough, I remember reading that he did the same experiment on a Conservative website with the Elephant techies, and though there were some minor differences in method to solutions, the substance wasn't as far apart as you might expect.

http://openleft.com (search the term "Durbin")

Thanks for doing this, I feel this issue is right up on top with the Iraq thing, since if the telecomms succeed in taking away our voice they've effectively killed democracy. And Neocons seem to want to allow them to do just that, quid pro quo for the NSA wiretaps.

Which leads me to make one more statement. Corporations need to be held accountable when they break the law. Don't let AT&T and other wiretappers off the hook.

3. *Senator John Kerry on High Speed Internet in America -- The Revival :: Independent Media and Politics* Says:
   September 23th, 2007 at 4:21 pm

   [...] join the discussion and leave us some comments. Thanks for visiting John Kerry is blogging over at SaveTheInternet.com. He hits the major points: broadband access is good for everyone, including small business. [...]

4. J-Ro Says:
   September 25th, 2007 at 4:33 pm

   Hi Senator Kerry!

   My idea would be to make the Internet more like a utility (gas, water, electricity, etc...). As time goes on, the Internet will only become more important to society. Even now, it is hard to fully participate in social and political life without reliable high speed Internet access. This problem will only get worse unless broadband access is recognized as a universally necessary service.

   The Internet as a utility would mean it is universally available, universally affordable, and completely net neutral.

   Thanks for listening. I look forward to hearing your proposals.

5. waynemanoele Says:
   September 25th, 2007 at 6:33 pm

   Senator Kerry,

   As I can see it, its clear that some sort of opening of access to the infrastructure that enables access to the Internet is necessary in addition to some sort of incentive for an increase in available infrastructure in those areas where it is unavailable.
   In my experience, the citizens of many communities are essentially left with one choice for broadband access if it is even available to them: DSL or Cable. Its also been my experience that in for the citizens of many communities, that limits you to a choice between one of two companies and whatever pricing schemes they may have available. To me, this points to a need for an open access infrastructure for Internet providing.
   Improving the infrastructure itself is tantamount it then seems. Having grown up in rural Minnesota, I'm quite familiar with the limitations of the current system we have. For many rural families, one is limited to dial-up, because a voice line is all that can reach their home. DSL can be limited distance, and if the cable company hasn't laid a cable line into your area, there's no access to a cable modem either.
   While the problem can be fairly easily defined, a solution is much more difficult. Forcing openness in communities with broadband access I suspect would be painfully difficult at best, and building whole new lines of access within the more sparsely populated areas of our nation would be quite expensive as well. To borrow the thinking of J-Ro above, perhaps the best solution would be to task a subset of our best engineers and scientists for whom this area is a specialty to work on creating a new infrastructure medium for the Internet that would be accessible much like a utility for all Americans. Given the difficulty of such a

task, in the interim we could press for the opening of the current infrastructure to alleviate the lack of access currently present.

Respectfully,
Wayne A. Marsalla
Eugene, OR

6. *Jy Says:*
    September 19th, 2007 at 7:08 pm

Jurgeno, I forgot to point out. You said, "As a resident of Boston, MA it is unbelievable that in this "world class" city we cannot get DSL, or FIOS into our neighborhood."

Right now Verizon is trying to kill Local Loop Unbundling in the last few places in the Northeast that still have some sort of Local Loop Unbundling in effect. Boston is one of those places, so it's going to get even worse if they succeed.

http://techdirt.com/the-fcc/tre-and-congress-you-gonna-listen-to-big-telecommedia-or-us

7. *J. Snow Says:*
    September 25th, 2007 at 7:43 pm

I was recently appalled to discover the 13 states have actually banned free wireless internet access. Meanwhile, the free wireless that is available in many metropolitan areas is barely functional at best.

Also, the functionality of US mobile internet devices lags well behind those in Japan, South Korea, and much of the EU, because US wireless technology is both considerably more expensive to use and less useful than in these nations.

8. *melson3 Says:*
    September 25th, 2007 at 7:54 pm

I agree with J-Ro, the internet should be supplied by a government utility that is highly regulated. This would provide inexpensive high-speed internet to all people. Often times, the big phone companies don't see a profit in building in some neighborhoods, so they just don't offer the service. This effectively cuts off everyone living in that area. Many rural areas seem to be in this predicament. It often takes non-profit organizations like Mountain Area Information Network in Western North Carolina to provide internet services.

https://www.main.nc.us/index.html

In addition to providing access to more people, Congress should pass a law that guarantees Net Neutrality as a fundamental right every American has. The internet should always facilitate and never impede the free flow of data and thus, ideas. Protect the Internet!

http://www.google.com/help/netneutrality.html

9. *KoreyAusTex Says:*
    September 25th, 2007 at 8:01 pm

I really like the idea that we make the internet into a utility, besides the airwaves are our property and all that goes with that. You know, we the people, and we should be able to decide who gets the contract to manage them on our BEHALF. Believe me the infrastructure argument does not hold water anymore and I am sure there are many companies that would drool over the thought of being able to help shape it's future. So treat it like a public utility and regulate it. I have a funny feeling that all those technological advancements they keep saying won't get implemented will get done if these corporations understand that this is OUR WAY OR THE HIGHWAY. I am not against them making profit but in this era of the DE-VALUED dollar those fat cats need to understand the days of passing their high costs of living on to the consumer are going to stop!

10. *JeffWazz Says:*
September 19th, 2007 at 8:36 pm

Senator Kerry,
True competition is the only thing mega corporation will listen to. At this point I think is should be the right of every citizen to have access to information that educates the populace. The government should run it (I agree with molson3) Just today, I found out I'm losing CPAN2 due to "digital" upgrading. Where's the FCC? Why don't we have net neutrality...and why oh why are the public airwaves used by corporations for corporations benefit and not the citizenry? If they effectively take out the internet, we will become a nation of lemmings...oops, did that already happen?

11. *eng2008 Says:*
September 19th, 2007 at 8:47 pm

The issues of freedom of information, informational equality, and business rights are at stake. Proponents of one side of the debate believe that all people have a fundamental right to information, and that no one should be denied that right by their internet service provider. The other side believes that businesses should have the right to regulate the flow of information over their service.
Because the government is in the business of ensuring a functioning market, and the rights of its citizens, there appears to be a dilemma. Fortunately for the Net Neutrality side, the case for freedom of information can win the debate from a market standpoint, as well as a human rights standpoint.
For the functioning of an ideal market, complete information is key. The more people with access to information, the more likely the market will produce an efficient level of goods and services. The internet is the perfect means to fulfill this need. It in fact has increased efficiency in the market dramatically since its inception.
The businesses who wish to regulate this information would impose far more costs on society than the benefits they would gain. Incredibly beneficial externalities that the internet provides act as a public good. The government has an incentive to protect those benefits for society and for the free market.

12. *James Says:*
September 19th, 2007 at 9:09 pm

Senator Kerry
I agree with bj. Verizon is trying to kill Local Loop Bundling. They are only selecting the areas they feel like coming into in our area. The competition that was promised is only in CERTAIN areas and mine is not one of them. So how is that competition. And Comcast keeps raising our rates. You only get the low rates for 1 year then you get thrown into a sort-of medium high rate. They give you a sort-of good deal. But Verizon is picking and choosing who they want to give their FIOS service to. Mostly into the rich neighborhoods and forced into the areas they were forced into so they could lay their fiber optic cable.

But we need to keep the high speed Internet available and competitive to everyone. I also like the utility idea. It would definitely bring the cost down. That is what we need. Even bundling has not brough the cost down. My neighborhood that I live in even having high speed does not get the benefits of the high speed. We are spposed to get 6 mbps. We barely get 2 mbps and 4 mbps tops for the year that I have had it and Comcast is still working on it. And I have been paying for high speed internet.

Please Senator, help keep the Internet alive and viable!!! Thank you

13. echrwnor Says:
September 29th, 2007 @ 8:12 pm

Internet access was once something extra, something a person could live without, a luxury. No more. It's now a necessity. So many transactions take place online — shopping, bill-paying, even renewing your car's registration and completing homework assignments — Internet access is something *everyone* must have.

How do you make sure the poorest among us have that access?

In addition to free access at libraries and other public places (e.g., city/town halls), free Wi-Fi access would help ensure everyone could gain access to the Internet. Whether this should be funded federally, statewide, or locally is the next discussion, but I think our goal should be free Wi-Fi access within our cities and towns.

Access should not be held ransom by a few big companies. It should not be private industry that decides who can and cannot get access, because of where they live, ability to pay, and the like. The ubiquity of everyday tasks to be completed online negates that.

Information wants to be free — and so does the Internet!

14. agnston Says:
September 29th, 2007 @ 8:14 pm

It is worth remembering that the Internet was developed mostly through federal funding (as a DARPA program, among others). As such, it can reasonably be regarded as a public utility. The privilege of using it should be won simply by paying taxes. As technical demands place an increased burden to provide speed, many citizens will begin to be excluded from the benefits this technological wonder provides. Telecomms want a two-tier system. One in which "premium" users have access to increased speed and other resources. They will claim they have to do this in order to cover the costs of improving existing infrastructure. At the same time, they will ask Congress for federal funds to help pay for it. This is a sure way to: a.) squander tax revenues (it wouldn't be the first time a Telecomm has squandered federal funds to help cover infrastructure needs), and b.) further widen the income/education/quality-of-life gap in this country.

I believe federal funds should indeed go toward improving Internet infrastructure and making it comparable to other developed nations (South Korea, for example). But if "we" pay for it, we should own it. If we end up with a truly enviable tool for the future, we should all share is its cost, and have equal access to its benefits. Ultimately, I believe this is the only way to ensure that all citizens have access to this increasingly important resource. What's true of public highways and public schools is true of the Internet - its too important to leave in private hands.

15. otbnalorty Says:
September 29th, 2007 @ 8:24 pm

Dear Senator John Kerry,

Everyone is giving an anecdote, so maybe an anecdote is a proper starting point. I moved out to California during the late 90s to join the workforce who were creating what we now see as the Internet. The atmosphere was too exuberant. Competition made it impossible for smaller fish to swim with Menlo Park backed IPOs sharks. We paid for this by losing many good companies that never came to fruition. The ironic aspect is that the Internet is constantly evolving. This evolution shouldn't be left to those with access to money, corporate friends, or access but to the dream that the Internet will look different tomorrow — that the Internet will belong to everyone tomorrow.

Back then, you could drive from East Palo Alto where funding for textbooks in classrooms wasn't dreamt of to Palo Alto where you saw a few people here and there throwing obsolete monitors into the trash. The connection is not made even today. The explorers who created the Internet of today knew about East Palo Alto and it sickened them.

I'll step off my soapbox. But before I do, I would like to say that Bush hasn't spread Democracy around the World, the Internet has. If access to the Internet (esp. high speed) you lose small businesses who most likely have solutions GE hasn't think up, you lose artists, musicians, people of conscience speaking out.

I have a belief that everyone possesses inalienable rights. The right to high speed Internet has become as inalienable as free speech. Am I the only one who still believes in rights?

16. _68brad Says:_
September 25th, 2007 at 9:36 pm

The attempt to privatize the Internet is but one more example of the legacy of taxpayer swindles that are the fetid legacy of the Reagan administration and the bipartisan antisocial mentality that has come in its wake. Even prior to this quarter century of extreme oppression, we witnessed near-monopolies of cable "service" whose unacknowledged leakage interfered with antenna reception and blackmailed even the most obstinate of non-customers into surrendering.

For all the propaganda that has been pounded into our heads for nearly a century, capitalism is not democracy—on the contrary, a dictatorship by corporate cronies, such as we have now, is fascism—and its mythical self-regulation does not render strict governmental control unnecessary, any more than the RICO laws are unnecessary.

17. _adr Says:_
September 25th, 2007 at 9:39 pm

Dear Senator Kerry:

In reading the other posts so far, a common theme appears to arise: competition.

It might have been the History Channel that ran a show a few years ago ranking the 100 most important inventions of all time. Ranked #1 was the printing press. Of course, the potential impact of the Internet on economic development, freedom and democracy worldwide makes the printing press seem quaint.

Why didn't television, a more widely distributed technology, outrank its centuries-old predecessor? Some might say it's because the television market was quickly consolidated by the then radio oligopolies, who were beholden to a similarly concentrated market of advertisers.

Senator, as you're probably well aware, our media market is now more concentrated than ever. Worse, the telecommunications dynasties are busier than ever doing all they can to close the Internet to competition for the sake of their already bloated executive paychecks.

There is simply no better champion democracy and the ideals of America's founders than a leader who would fight to safeguard the freedom this new technology can usher in. And such protection requires no taxpayer dollars:

Step 1 - Pass a "net-neutrality" act, guaranteeing equal access to the Internet and preventing the "televisionization" of this new engine of democracy and economic growth.

Step 2 - Pass another no-taxes-required bill preventing the telecomm. companies from outlawing municipal and other wireless ISP start-ups, as they did in Pennsylvania.

Step 3 - The heck with it. If you accomplish steps 1 and 2, you'll already be a hero.

Thanks for inviting these comments. Keep fighting the good fight.

Demorate for Democracy,
New York

18. *Ikarr Says:*
September 29th, 2007 at 8:40 pm

Dear Sen. Kerry,

My question for you is: Are you willing to take a brave stand against special interests, to listen instead to the public and open the Internet for everyone?

Everybody — with the possible exception of the phone and cable companies — seems to agree that connecting more people to a faster, open and more affordable Internet is a good thing. The question is: how?

The powerful Telco lobby would have us believe that a "hands off" approach is best — otherwise known as "hand over" control of the Internet to them.

We have done this already — lobbyists have strong-armed Washington into ceding to the phone and cable companies near complete command of the market. Today this duopoly controls more than 96% of residential broadband connections. But what has this got us? A failed marketplace, especially when compared to what other advanced countries now offer their citizens.

The best way to restore America's prominence is to have Congresspeople like you take a stand against the telcos and ensure that America's communications infrastructure benefits the common good. Like the public highways, the information superhighway must be considered a key piece of public infrastructure — an indispensable part of our society that provides vast economic and social benefits to all.

To that end, it's important to support bills such as the "Community Broadband Act of 2007," which allows cities to wire their citizens much like a 21st Century library system.

It's also important to see that the FCC allows unlicensed access to "white spaces" to open up new possibilities for the next generation of mobile Internet devices and municipal networks.

Brave leadership involves seeing that all broadband networks — whether wired or wireless — are open to all producers and consumers of Internet content on fair and equal terms without discrimination. "Net Neutrality" and "Open Access" offer maximum choice to all consumers while fostering competition and innovation where it is needed most. It also stops self-interested gatekeepers from holding captive the online marketplace of ideas.

It's important to support the "Internet Freedom Preservation Act," which protects Net Neutrality and stops gatekeepers from blocking, degrading or slowing down content they don't like.

America must look at our Internet as one of the great public works projects of our time and build the world's most advanced communications networks without stifling the free flow of information that has made it so important.

As an influential senator, You can take the lead to make an Internet that's for everyone. Compromise at nothing less.

19. *JJ Says:*
September 29th, 2007 at 9:48 pm

As a person with disabilities the internet is a godsend, but in the form of dial up which is becoming neglected by all companies as they want one to pay the out of sight monthly fee for high speed. In Rural America and in this town we have but one option for either service with No Competition. WE cannot join high speed at a reasonable price ' it is not within the average budget of a person on a fixed income.. Please someone use come common sense and not allow the big Corporations to take away this control and allow the free enterprise system flow free for everyone and without being with one provider it is unfair costly and not in the American way of doing things ..
Profits for Stockholders and Corporation salaries come first then the Consumer . this internet is the lifeline to the the world for all and especially those who cannot explore much of the world in person .
So let's let democracy work and not allow such restricted access especially in Rural America... Monopolies are not warranted for a internet the taxpayers invested in long ago and now BIG BROTHER Corporations want to control all at a drastic price ~ Please don't buy into this rape of the public voice across the country and others.

20. *reinstate Says:*
September 29th, 2007 at 9:47 pm

Hello, Senator Kerry,

It is the size of the dinosaur that determines who gets the bones of high speed Internet connectivity. Here is a specific example from rural Middle Tennessee where I live.

BellSouth, now AT&T again, has been the sole telephone service provider in my area - they have a monopoly on local service connections which also includes Internet connections. In the largest town in my county, there is DSL in most of the city and there is also competition from Comcast cable. The rest of the county has no high speed access from anyone. Why? Not enough profit for AT&T. It has been this way since I lived here the past 10 years.

In the adjacent county, one of the least affluent and least populated in Middle Tennessee, ANYONE WHO WANTS to pay for it can get high speed DSL. Why? Because the county and several others adjacent counties have as their prime telephone service provider a rural telephone cooperative which is NOT BellSouth, not AT&T. The board members of the cooperative decided 6 years ago that they were going to provide their service area with DSL and they covered their three county area in less than two years.

When the government decided to let AT&T re-glomerate, that was one of the worst decisions ever. AT&T still holds monopoly power over most of their service area, especially in rural Tennessee, and unless they are forced to provide the public services required of monopoly utilities, they will do as they damned well please.

If you want to see AT&T move, then pull the plug on their monopoly and allow other service providers to use their lines FOR FREE to provide high speed services. They have plenty of bandwidth. You will be amazed at how fast a dinosaur can move.

Cheers,

21. *Josh Says:*
September 29th, 2007 at 4:50 pm

We are being held hostage. Were's home land security?? AT&T and their cronies are capable of providing the best internet structure in the world. But not until they are allowed to charge a arm and a leg for it. They want you to know that it costs money to do this. And they have the Senator in my home state pounding the drum for them (Gordon H Smith). But the truth is if they would make it more affordable to the masses, they would be making more money than the billions they're making now. Read my story
http://www.savetheinternet.com/yourstory/295832

22. *DeanSB2000 Says:*
September 19th, 2007 at 8:51 pm

I'm one of the few lucky ones, because Storm Lake, Iowa has 2 cable TV, 2 cable-modem high-speed Internet, 1 DSL high-speed Internet, and 3 phone companies to choose from.

But if you live in most other areas of Iowa, you're stuck with either Mediacom for high-speed Internet access through cable, or Qwest for high-speed access via DSL.

That is called a DUOPOLY, and BOTH providers want to be given the ability to become "gatekeepers" who would THROTTLE ANY traffic that doesn't have a "sweetheart deal" with them.

That means that they would have the ability to decide, arbitrarily, WHICH websites you could access at high speed, which ones you could only access at slow speed, and which sites wouldn't load up AT ALL, because the providers decided to either outright BLOCK access to those sites, or THROTTLE access to them!!

That is why, until there are MORE competitors in EVERY neighborhood in this country, regardless of whether or not they're rich, poor, or middle-class, the U.S. MUST have ENFORCEABLE Network Neutrality regulations on the books, to PREVENT these "incumbent" providers from turning the Internet into a "two-tier" system of "haves" and "have-nots"!!

So, PLEASE put REAL Network Neutrality regulations through Congress...PUSH 'til it GIVES!!

DON'T let cable and phone company lobbyists decide the Internet's fate for the future!! PLEASE let there be MORE competition, and a NEUTRAL, OPEN, and FREE Internet FOR ALL!!

23. ████████████████ Says:
September 20th, 2007 at 9:54 pm

Thanks for asking!

Please consider the importance of Public, Education, and Government (PEG) services that need access to broadband infrastructure. We need your help to support non-commercial voices and ensure Localism and Diversity in our evolving communications sector.

The concept of PEG Broadband is a useful way to organize the fundamental public policy questions that need to be addressed. As you consider the issues for next generation digital media, there are many important lessons learned from last generation analog media... here's a modified summary of comments offered to the FCC's at a Localism hearing back in 2004 http://www.fcc.gov/localism/hearing-monterey072104.html

1. Commercial media/telecom alone do not adequately serve local community needs and interests, and consolidated ownership exacerbates the problem.
2. Local public interests are at stake as Congress and the FCC reshape the regulatory landscape.
3. The best way to promote PEG Broadband services is to ensure local and diverse ownership and to set aside bandwidth with adequate operating support for non-commercial, public service media/communications in every local community.
4. PEG access media provide a model for PEG Broadband to support true localism in our media/communications sector.

5. Local and State governments must have meaningful roles to adequately protect consumers and to effectively advocate for local needs and interests to be met.

Thanks for your thoughtful consideration of PEG Broadband!

24. AnEvil Says:
September 28th, 2007 at 9:01 pm

First off, I think it's great a large public figure is taking on this effort head on. The current broadband situation in the United States is ridiculous and reeks of fraud, corrupt business practice, and more than ever - the placement of special interest group needs over that of the general public.

I've been following the Comcast broadband issue ever-so-closely over the last few months with their recent P2P crackdown (which I find disheartening considering it's legitimate usage). This website (more or less) sums up my frustration: http://comcastissue.blogspot.com/

Now will you read it STI? Will John Kerry read it in it's entirety? I can't help but feel this issue by a public congressmen will off the bat be poorly represented because of a lack of understanding/education on the subject. The ideal leader of this fight would have a degree in Computer Information Science and have respective knowledge of network deployment, scalability, and management. Notwithstanding the lack of a computer literate government officials, I don't think all is lost.

Government needs to first off step in and take control of the infrastructure. Government SHOULD lay the lines, maintain them, and prevent blackouts - making sure that content from the various datacenters around the nation are peered correctly. That's it. To inspire competition, providers should "provide" service and consumers should be able to pick from a wide variety of options. I don't understand why monopolistic competition is allowed to run it's course in our country. This argument could inevitably take me down a whole entire different path of discussion of domestic priority and the shortcomings of our current administration. But I digress....

Overall, this problem is multi-faceted. Special interest groups and lobbyists need to be put in their place. Government needs to step up and start regulating unfair business practices and take control of the infrastructure (and upgrade it). Legislation needs to be re-written and/or revised - you all realize that the current FCC definition of Broadband is anything above 200K. Maybe I threw away my 14.4k modem a bit early...

Keep fighting and let people hear your words. They can't ignore you forever.

25. madhspawn Says:
September 30th, 2007 at 9:04 pm

Aside from protecting net neutrality, I would investigate the $200 billion theft by major phone companies when they failed to complete contracts for upgrading internet infrastructure in certain areas of the country:

http://www.pbs.org/cringely/pulpit/2007/pulpit_20070810_002983.html

If we're serious about improving the infrastructure in this country, we can't let big companies rip us off without delivering anything. Not only will this not improve the country's broadband capabilities, it will also encourage companies to steal money for future projects that may come of this discussion.

26. aWVine Says:
September 28th, 2007 at 9:06 pm

I spend part of the year in Paris, France where all the phone and cable companies (both privately owned and state owned) are offering a complete package of DSL/telephone/basic cable TV for about $20 a month plus tax. What makes it even better is a free modem and a free decoder box for the TV. If that's not enough all phone calls are toll free – to France, to Europe, to North Africa, to North America and a bunch of other places as well. As a result, the only extra charges are for premium TV channels. Why can't we have this here?

27. *Jeff Says:*
September 25th, 2007 at 9:19 pm

THE INTERNET SHOULD BE LIKE RADIO, FREE TO ALL WHO TUNE IN. IT SHOULD "PLAY" AT THE SAME SPEED FOR ALL. THE INFORMATION SOCIETY CANNOT HAVE SOME WITH ACCESS AND SOME WITHOUT....THAT HAS NEVER BEEN BENEFICIAL. AT THE SPEED CHANGES ARE OCCURING, THERE IS NO POINT IN NOT COMBINING TV, PHONE AND INTERNET INTO A ONE STOP SERVICE AVAILABLE FROM MANY, MANY PROVIDERS. THERE IS NO LOST PROFIT IN FREE ACCESS TO CONSUMERS!!

28. *occupier Says:*
September 25th, 2007 at 9:25 pm

Telephone companies such as AT&T and Comcast charge a small fortune for their services. I am not sure that the government should actually own the lines, as then repairs would really be bogged down. When they broke up Ma Bell it was the best thing that could have happened. Now the conglomerates are back; what happened to the laws about monopolies? The internet should be readily available to anyone who needs it, not just those areas whom the big companies are willing to place lines in. The internet should be combined infrastructure now and maintained as such. The lines should be laid and serviced with an eye to everyone, not just a few, and should NOT be in the control of a few. There are some excellent ideas on this forum, and for once I would like to see a politician put his money where his mouth is and work for someone else besides himself, as far as I am concerned there is not a single politician in this country that is truly representing US; they mostly are in it for themselves and we are the losers. Kudos to you if you stand up for what is right; we, the voters, are watching closely what happens in this issue. Don't even get me started on Cable TV, where you have to pay for THEIR choices! The nonsense has got to stop. The internet cannot be controlled by a few who wish to limit access. It is a wonderful thing and having gatekeepers who decide who gets what would ruin this. The gatekeepers wannabees did not invent this and should not be in control. Period, Case Closed, the internet belongs to everyone. It is time we caught up with the rest of the world. Out with the dinosaurs.

29. *b-muffin Says:*
September 25th, 2007 at 5:21 pm

I think it is essential for our country to start considering broadband a public utility. However, there are a few things to keep in mind with how this plays out in reality.

Municipalities are not always able to take on the building of a reliable broadband infrastructure on their own, and are forced to look to the private sector for help with funding for construction, maintenance, and supplying internet service.

Instead of forcing them to take private money and sign binding contracts that are not always in the public interest (and will lead to redlining, price increases, etc.), money for these projects needs to come from other public outlets.

Beyond the baseline of passing federal legislation to override state laws that currently ban publicly-owned broadband networks, let's also propose federal funding for municipal broadband projects, at least until they get off the ground or a list of reliable implementation models becomes available for municipal providers.

Already, many many cities and towns around the country have implemented municipal broadband. But it seems to me that the success of these projects often relies on some more or less must-haves (smaller population size, flatter topography, profoundly no competition from a stronger private telecom, etc). Especially in large cities (New York, San Francisco, etc), governments can't even rely on private money to help with build out, because companies look for the kind of profits that seem virtually impossible to achieve at this point from a free or low-cost broadband network (even wireless, not to mention wired broadband). These cities are left with no options for publicly-owned, city-wide broadband as of today.

The point of this ramble is that it is not enough to 'consider' broadband a public utility — we must take into consideration current information available on how cities have tried (and succeeded or failed) to make it so, and based on that information, help them financially to build out these networks.

If municipal broadband succeeds, it will offer strong competition to incumbent telecom providers, forcing buildout and lowering prices. But it is an infant 'industry' with a lot of big dogs already in the market. The federal government needs to allocate money to help it grow.

---

30. okraus Says:
September 29th, 2007 at 9:24 pm

Dear Senator Kerry,

This is another in a long line of sad commentaries where corporate interests outweigh the needs or the rights of the people in the USA. Night after night, issue after issue, our news media reports how the people's voice is ignored in Washington, D.C.

Thanks for taking up this important issue. Please pay heed to the wise comments from everyone here who has taken the time to post a response in this blog.

It is important to note that competition is a key theme to these responses and a necessary ingredient in any antidote to this situation. Internet Neutrality is a must. In an effort for the USA to remain competitive in the world economy and its people to be first class internet citizens, we hope you are successful in garnering support for the Internet Freedom Preservation Act.

Best of luck in your efforts.

---

31. heresmypoint Says:
September 29th, 2007 at 9:33 pm

Senator Kerry,

I would please ask that you support a free Internet, or 'Net Neutrality'. There are many programs and options available to people and organizations who feel the need to filter their web searches, protect children from browsing potentially harmful sites or prevent employees from accessing 'banned' sites. We as individuals should be allowed to choose to filter, or not to filter, our own home based internet connections. I was surprised to learn my own connection is being filtered. Sadly, in my area, I have no other viable options for accessing the internet.

---

32. Rexcredletstein Says:
September 29th, 2007 at 9:32 am

Let us get a quick grip on this issue. India and China will be passing us by and lots of other lesser nations out pace our broadband speed and usage already. What is the excuse...government and the telcos are in bed together. Competition, regulation, both...and fast action is needed. I live in a rural area of Northern California. We are severely underserved by CATV and Telcos. There is no incentive or desire by the providers to stretch out to missed customers who can only resort to slow dial up or maybe slow and expensive satellite. Let's redirect some of those telco lobbying dollars to solving this problem.

---

33. oraclesean Says:
September 29th, 2007 at 9:41 pm

Senator Kerry,

Thank you for participating in this forum. This is how our democracy should be, where individuals have an equal voice to lobbyists, and money doesn't let any single group bend the ears of lawmakers more than common citizens.

As the CEO of a small technology consulting business, I rely on the Internet. Almost all of my work is conducted remotely. I am fortunate to be in an area that was recently enabled to use FIOS, and the speed is excellent. Obtaining the service and dealing with their customer service is a trial in patience, but what do you expect from a monopoly (Verizon)?

My primary concern is the desire of telecom companies to throttle service based on who you are or what you're willing to pay. I should not have to pay more to get preferential treatment. I should not have to pay more to avoid having my web site blocked by an ISP. The mob did/does that, and it's called a protection racket.

The Internet, like telephone service, should be equal to all. No service provider should be in a position to censor, limit, or exclude content it serves to its customers.

I know that my situation is not unique. I'm sure that the Verizon's and AT&T's of the world would be pleased as punch to collect the extra revenue, but the result would be devastating to businesses such as my own. We must have a Net Neutrality initiative in place before we take any further action to fix the Internet.

If legitimate competition existed, ISPs wouldn't dream of taking away services, and they'd be falling over themselves to offer cheaper, faster, better service. Look at the cellular phone market, where you can freely switch service providers. Costs are affordable, packages are varied, and innovations in service are high. Contrast that to Internet providers, where choices are limited at best, and we continue to fall farther and farther behind.

As a side benefit, an open, neutral net promises to make us a greener, healthier, and happier nation. My company provides specialized services to customers all over the country, but we all work either from home or a local office. I no longer commute, so my car isn't on the road, contributing to the problem of global warming and traffic congestion. I'm home for my family most of the time, which makes the bond with my wife and children stronger. No longer do I leave in the morning before they're awake, and return home in time to tuck them in. I'm healthier, because the time that used to be spent commuting, I now spend cycling or roller blading with my daughters. I'm not breathing contaminated, recycled air all day, and my productivity is much higher.

My business is not unique. As our economy shifts to being more technologically driven, more and more employees will be in a position to work from home at least part of the week. Happy, healthy employees are more productive, and make us a stronger nation, but it can't happen without competition in the ISP market and enforcible Net Neutrality laws.

Thank you for your time and concern for this issue!

34. Griffon Says:
September 20th, 2007 at 3:41 pm

It seems to me that the Internet is still such a new technology, that much regulation at all with regard to providing it is liable to stifle new access much more than it is likely to help it. How can you help something grow if you don't know what nourishes it?

Congress did a very smart thing in the early days of the Internet in refraining from over-regulating the content of the Web, and the type of commerce that went on. As a result it flourished.

Unfortunately, they took the opposite approach when it came to the physical infrastructure that made up the internet. They allocated a bunch of money for very specific purposes, and gave it to a only a few entities. How is that supposed to spark innovation? The result is that most companies STILL run on T1 connections that, if we are lucky, only cost us half of what they did 10 years ago. While this may seem like progress to some, in the technology world, things double in power or halve in cost every 6 months. This is stagnation in the worst sense.

Big old companies have little to gain from innovation. It's the small companies that drive innovation and development. If you want innovation, you have to help the little guys. The big guys will figure out that they need to innovate or die – and believe it or not, they will find the money to pay for it from their own pockets if they have to. Stop protecting existing providers. Offer rewards for results, not contracts for plans. Look at the X-prize.

Oh, and as a general rule, stop trying to regulate technology. The last thing we need is people making up rules for things they do not understand. "The Internet is a series of tubes." Good grief! I don't mind a little regulation in context, but when someone suggests that there might be a completely new use for a piece of spectrum, why on earth would you not let them try to make something of it?

35. *TimELiebe* Says:
September 29th, 2007 at 9:53 pm

More than anything, I think you and every every Democrat needs to get behind Net Neutrality – and STAY behind it, no matter what the Republicans allege. It's vitally important to educate the American public that the Internet belongs to THEM (as has been pointed out elsewhere in this thread, the Internet was originally considered part of our country's defense, and should have been made a public utility) – and not to media conglomerates like Time Warner, Verizon and Cox.

If Big Media is allowed to have their way, the U.S. will become a second-class country where innovation and creativity is stifled, or only permitted in a very limited sense if the Corporate Masters PERMIT it.

36. *kookhans* Says:
September 29th, 2007 at 9:55 pm

The internet is the technology of now. And broadband should be open to all Americans in America. I have some friends one in VA and the other in New Mexico and they can not get broadband.
I am glad I can but, this is something that should not be optional and only have dial-up service. I am speaking those they did a voice.

37. *madreaNle* Says:
September 29th, 2007 at 9:53 pm

Limiting service options to cable, satellite or [monopolized] telephone lines has effectively limited public access to the internet. This has deleterious results no: business, education and an informed citizenry. While in the short term, a regime that controls communications consolidates power in such manner, in the long term these ill-gotten "gains" serve to diminish these resources..
Strength through diversity, regulation to stop monopoly, protection of freedom to choose sources, REAL security issues. Please weigh in positively for "Net Neutrality" a soubriquet indeed for "Internet Freedom" but what we need by any name.

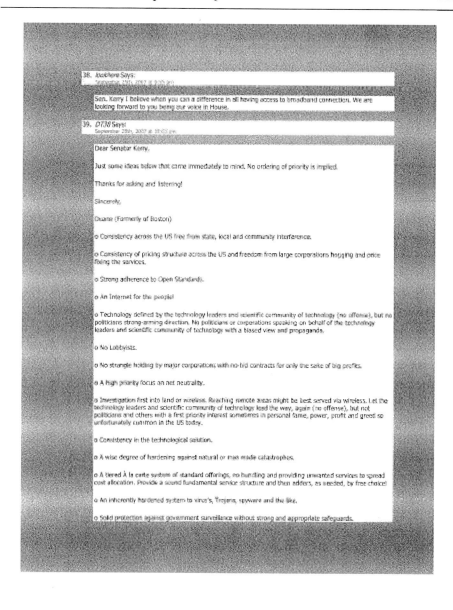

38. *JackFerd Says:*
September 29th, 2007 at 2:05 pm

Sen. Kerry I believe when you can a difference in all having access to broadband connection. We are looking forward to you being our voice in House.

39. *D736 Says:*
September 29th, 2007 at 10:03 pm

Dear Senator Kerry,

Just some ideas below that came immediately to mind. No ordering of priority is implied.

Thanks for asking and listening!

Sincerely,

Duane (Formerly of Boston)

o Consistency across the US free from state, local and community interference.

o Consistency of pricing structure across the US and freedom from large corporations hogging and price fixing the services.

o Strong adherence to Open Standards.

o An Internet for the people!

o Technology defined by the technology leaders and scientific community of technology (no offense), but no politicians strong-arming direction. No politicians or corporations speaking on behalf of the technology leaders and scientific community of technology with a biased view and propaganda.

o No Lobbyists.

o No strangle holding by major corporations with no-bid contracts for only the sake of big profits.

o A high priority focus on net neutrality.

o Investigation first into land or wireless. Reaching remote areas might be best served via wireless. Let the technology leaders and scientific community of technology lead the way, again (no offense), but not politicians and others with a first priority interest sometimes in personal fame, power, profit and greed so unfortunately common in the US today.

o Consistency in the technological solution.

o A wise degree of hardening against natural or man made catastrophes.

o A tiered à la carte system of standard offerings, no bundling and providing unwanted services to spread cost allocation. Provide a sound fundamental service structure and then addons, as needed, by free choice!

o An inherently hardened system to virus's, Trojans, spyware and the like.

o Solid protection against government surveillance without strong and appropriate safeguards.

o Guaranteed speed across the US. Broadband should be the top priority.

Note: I have a guaranteed speed of 5mbit, soon to be 8mbit 7×24 from https://everestkc.net/ and that is what I get. Others should be able to do this, eventually. We need to stop consistently falling behind in the US.

40. *barrison* Says:
September 28th, 2007 at 10:04 pm

I am an applications developer in New York City. I didn't study computer science in school but sort of fell into the business during the dotcom boom in 2000. Who knew that a struggling musician would love programming so much? With that said, I am concerned that if our country doesn't provide cheap/free, easy, wireless access to the internet, unfettered by corporate greed, as well as improving our public school system we will be technologically surpassed by other countries that do invest in this type of infrastructure. We must remain strong and competitive and I fear if we fail to do so in search of short term corporate profits we do so at our own peril.

41. *neodoc* Says:
September 29th, 2007 at 18:05 pm

I'm surprised, aside from the obvious comments about monopolistic control of the Internet via access - that no one has pointed out that universal broadband access is not only possible - but feasible now. Companies like Telkonet can supply broadband access via electrical lines that are connected to every home in the country. Why this hasn't taken off is beyond me - aside from the political will that is absent (lobbyists?). The same argument holds for solar power - every home and business that has a roof could contribute to a solar grid that would provide the US with all the power it needs! Gee - what is stopping us? (lobbyists?). What is needed in this country - is a new revolution - to free us from greedy corporations and weak-kneed politicians that can only emit a lot of hot air but no light!

42. *Brad7* Says:
September 29th, 2007 at 10:18 pm

I have to agree with what everyone above is saying. The internet should be considered an utility like electricity, so that everyone can have access to to it at fair prices.

43. *Silent Lucidity » $200 Billion Rip-Off* Says:
September 29th, 2007 at 10:17 pm

[...] is in regard to a guest blog post by Sen. John Kerry: Questions from Sen. Kerry: How Can We Connect America?, sent as a trackback.] Technorati Tags: broadband, Internet, John Kerry, net [...]

44. *DanielZimmerman* Says:
September 29th, 2007 at 10:23 pm

Senator Kerry,

I agree with "adz," and I believe in an open Internet, free from monopolistic control of territories by private enterprises. I also believe that the Internet should function as a way for citizens to vote on public referenda affecting the entire country. For example: four issues per year affecting everyone could open for discussion on a dedicated web site (the involvement of cable television would help immensely in this endeavor), with two months open to discussion based on peer-reviewed data, and one month open for people to vote on the issue. (As an academic, I would prefer that voters would have to qualify to vote by taking an online test of their knowledge of the issue, rather than merely "vote their gut.") The site might require a certain percentage of votes (say, a 60% majority) in order for the vote to count—and "count" would mean that when the issue comes before Congress, the vote of the participants (and Congress might stipulate a minimum number to qualify) would count as a fixed percentage of the House's or Senate's vote—for example, 5% or 10%—a modest but potentially decisive voice in national decision-making. We have entered an age that permits direct participatory democracy. This proposal would require an amendment to the Constitution, but one that would encourage an electorate so alienated from the political process that barely half of eligible voters exercise their franchise to care again about their country. It could make the people directly responsible for the kind of world propose to live in. It could make the Internet and related communication outlets an integral part of the country even more than it has become an integral part of the economy. Such a proposal requires the same kind of universal access as citizens have to the voting booth (recognizing stipulations like age which restrict, e.g., minors from voting). That kind of access, in turn, requires significant modifications in the present configuration of Internet access and related legislation. The people of the United States need a meaningful voice in their own governance, and present technology offers the opportunity to make that voice decisively heard.

Sincerely,
Daniel Zimmerman

45. *blackmachead07* Says:
September 25th, 2007 at 19:36 pm

Greetings Senator Kerry!

Please consider the importance of ensuring that every America having access to high speed connectivity to the Internet!

As you may be aware, the Internet has become one of the most important communications tool across the entire world. However, here in America, we have allowed mega corporations to decide what type of Internet access the American people will have. For too long, conglomerates such as; Verizon, AOL, ATT, Comcast and other Internet Service Providers (ISPs) have controlled the speed at which we transmit data from one datapoint to another. They have done so, in a deliberate manner to fatten their pockets and we, the people must be given a chance to add balance to this important debate.

As a Senator in the 110th Congress, you have the power to ensure that American citizens will not be sold out and blocked out from having equal access to high speed service over the Internet. Which is by the way, made abundantly available all across the world, with the exception of the United States of America.

I strongly encourage you to do whatever is necessary to provide the citizens of this great country fair and equitable access to high speed Internet service to every locale in these United States.

Don't let the conglomerates write their own bills that will guarantee them great wealth for generations to come. We the people deserve to have laws that will be fair and balanced across the Internet speed spectrum. I don't think it would be fair to allow ISPs to segment various levels of speed to the Internet for a premium level price. When in reality, it does not cost corporations any more or any loss to open up high speed Internet access to everyone. When in fact, the slicing and dicing up the various access speed levels to the Internet will in of itself create more overhead costs in the long run.

If other countries can provide high speed service to the Internet for a much cheaper price, then why can't the American people have the same level of access too faster service for lower costs?

Afterall, ISPs are using public right-of-ways to install the FIFOS infrastructure and have done so by using cheap labor (illegal immigrants) to perform these tasks. I have personally seen this happened within my own community. And within days after the lines have been installed, I received hand delivered brochures that were hung on my doorknob from the companies who solicit high speed access to the Internet for a low introductory price of $99.00 per month. Out of that $99.00 per month premium, I estimate the company's profit margin to be around 45%-60% per household(conservatively speaking). Now, I interpret this as intentional highway robbery! More importantly, it is exploitation of the workers and to citizens they are serving.

Therefore, I strongly emphasize that you take a bold step in being the first Senator to support legislature that will indeed be fair and balanced toward the PEOPLE!

In closing, I would like to leave you with this verse:

The president is up there in the White House for you...not you here for him. The secretaries act in their bureaus for you...not you here for them. The Congress convenes every December for you. Laws, courts, the forming of states, the charter of cities, the going and coming of commerce and mails are all for you.
~Walt Whitman

Thank you.

---

45. *MonkaHouston Says:*
September 20th, 2007 at 10:22 pm

Senator Kerry,

I am encouraged by your effort to listen to the average citizen when it comes to this issue. America was once on the forefront of communication technology in both development/research and in deployment. We have, unforgivably, fallen behind many other countries in this regard.

We have reached a point in this country where a few powerful organizations have far too much say in the development of laws that work against bringing the necessary Internet services to the average American. With the existing technology, we have the means by which to provide Internet access to almost every American. The use of frequencies being released as we move towards DTV could result in America being the first country to make a monumental step towards bridging the digital divide. Imagine a country where the majority of citizens have free access to information. The potential for economic growth by providing this type of access is tremendous. This is only one method to bring universal broadband access to everyone.

Additionally, regardless of what others say, consumer choice with regard to high-speed Internet access is a joke. I currently live in Houston, TX within walking distance of the largest medical center in the world and I only have two companies to choose from when it comes to high speed internet, Comcast Cable or AT&T. Both of these companies offer plans similar to each other but not on par with other countries and neither uses fiber optic to the house. They may use fiber to an access point in the neighborhood, but without fiber to the house, internet speeds will never rival those in other countries.

By the way, why hasn't AT&T or any other telecom company been held liable for the billions in breaks they received to run fiber throughout the country since they never actually did it?

Another important issue facing America is the issue of providing regulations regarding Net Neutrality. As you know, this is an issue that cannot be ignored. Without any regulations to guarantee unrestricted access to the entire Internet (not just the parts pre-selected by corporations) the country would be at a major loss and regardless of the lobbyists statements, innovation and development would be crippled.

Again, I applaud that you are willing to continue to research this issue outside of the telecomm lobbyists views and hope to see promising action in congress by you and other representatives.

47. *bridgette* Says:
September 25th, 2007 at 10:35 pm

Hello, Senator Kerry,

Thank you very much for paying serious attention to this crucial issue and thank you even more for giving us an opportunity to comment.

I'm a poster child for the problem. My husband and I have an internet based business. We live in an idyllic, fairly rural environment which is, of course, seriously underserved by the telecommunications industry. There is no DSL available, the cable lines stop about 2 miles away. Dial up on old phone lines doesn't allow for any of the functions we need to maintain our business. (Have you ever tried uploading photos of products to a website on dialup on old phone lines?)

To have access to broadband so we can run our business, we had to invest in an expensive satellite dish and pay more in monthly charges than our friends 3 miles away who have DSL. We are grateful that we at least have an option but that we should have to pay this much more for a service that people 3 miles away have easy access to is insane.

The problem with satellite broadband (in addition to the cost) is that useage is limited to a certain amount of megs/24 hour period. If you exceed that, your internet connection drops into the slow lane for the next 24 hours. We've resorted to driving into town to the library with our laptops to download software updates and we can't take advantage of much of what's available on the internet in video and still have enough bandwidth left to run our business.

This is just insane! Access to the internet today is like access to electricity in my great grandmother's time. The internet should be classified as a public utility in the same way as electricity and water are because, increasingly, it is essential to survival in this global economy. Japan and Europe are light years ahead of us with internet access and their economies and societies are thriving because of it. The US will not be able to compete in a global economy without major changes in access to the internet.

Thank you, again, for being an activist for this issue. Please do what you can to help us.

48. *meeoohloo* Says:
September 25th, 2007 at 11:54 pm

We are losing, or have lost, so many of our rights under the constitution (which I now must type in lower case as it's been abused so much by the Republican administration) and have so little left. Our media is owned and operated by greedy corporations. This was supposed to be a country of individuals, supported by a strong Constitution, bolstered by education for all, with a free press and a strong representative government. At times it seems as though all we have left is the internet. Protecting the internet is crucial as it has become the last bastion of freedom in this country. Please do all you can to protect our rights.

49. *Mark H2* Says:
September 25th, 2007 at 11:52 pm

Senator Kerry,

**From: The Washington Post - washingtonpost.com

http://tinyurl.com/3rxpt5

"Japan's Warp-Speed Ride to Internet Future"

"By Blaine Harden
Washington Post Foreign Service
Wednesday, August 29, 2007; Page A01"

"TOKYO — Americans invented the Internet, but the Japanese are running away with it."

"Broadband service here is eight to 30 times as fast as in the United States — and considerably cheaper. Japan has the world's fastest Internet connections, delivering more data at a lower cost than anywhere else, recent studies show."

**To find out why, read the article.

**More:

"In 2000, the Japanese government seized its advantage in wire. In sharp contrast to the Bush administration over the same time period, regulators here compelled big phone companies to open up wires to upstart Internet providers."

**COMPELLED BIG PHONE COMPANIES TO OPEN UP WIRES TO UPSTART INTERNET PROVIDERS!!!

**Competition is key. Competition is THE key.

**I've said enough, the article speaks for itself.

**Thank you, and respectfully,

**Mark Holt

50. barccornick Says:
September 25th, 2007 at 11:08 pm

I can't imagine life without the internet. I use it daily. It's more important to me than any other means of communication. I can find information I trust there, I can find out what is going on in the world without being assaulted by blow dried, face lifted nitwits spouting inanities. Once it was against the law for any corporation or person to own more than a small portion of communication in this country - since Ronald Reagan destroyed the anti-trust legislation that protected our right to know - and made it possible for someone like Rupert Murdoch (who has made a fortune combining explicit sex with conservative politics in every possible media), we've come to depend almost exclusively on the internet. This country is based on knowledge and education. It's a crime that we only have one means of communication where we can find out what's real in the world - but since that's what we've got, then we must keep it intact and as much as possible out of the hands of the greedy wretches who want to control our information and thus our thoughts. This may be THE modern issue, as it could make or break every other issue that is on the table nationally.

51. Joegam Says:
September 25th, 2007 at 11:09 pm

Dear Senator Kerry,

Please support net neutrality. As all the mainstream media become more and more centralized, the people... and the republic need the diversity of information and views that the internet provides.

Thank you,
-Joe Gumbosky

52. mttfc Says:
September 29th, 2007 at 11:28 pm

Dear Senator Kerry,

Thank you for providing this opportunity to address the issue of broadband service in the US.

First, we need not neutrality.

Second, we need to put some effort into how we deliver broadband without conflicting services reducing the value of the purchased broadband. In my home, it isn't a problem, but in my sons' apartment where they are paying for the broadband access, the competing services can effectively negate the value of what they are paying for. I am of the opinion that broadband access is like the old telephone access and improved by monopoly and regulation, not by free competition. I spent a weekend with 2 of my sons during which they could not use the broadband access they paid for because someone with a competing service was blanking their access channels.

I am not a technician, just an economist. Sometimes monopoly rights with appropriate regulation work better than the free market.

Sincerely,
Mick Cox
San Diego CA

53. Bud Murphy Says:
September 29th, 2007 at 11:25 pm

When it comes to the Internet and high speed access Senator Kerry's question "how can we connect?" is not the important question. We have the technology to provide universal affordable access to the Internet. The more fundamental question is, are the "shock economists" willing to invest in the future or are they blinded by short sighted economic models that see quarterly profits as the only desirable goal?

54. Teddy3Bears Says:
September 29th, 2007 at 11:33 pm

Dear Senator Kerry,

As CFO of a company that is seeking to exploit the accessibility of the Internet, I do have some things to say.

First, as you know none of the major Telco's or Inbound ISP's have invested one thin dime in the research and development of the Internet, this has been solely funded by the American taxpayers as the defense project called ARPANet. Which was designed for one purpose, to provide the United States with a highly resilient data-communications network that would be nearly impossible to destroy by an enemy. Their investments have only been in the hardware necessary to provide the infrastructure from point to point. Even this has been offset by the fact that MaBell had already invested in telephone poles and underground conduits that are used, for which we the taxpayers also subsidized with granting them a monopoly within the community. So far, we the taxpayers have shared the costs for companies to provide telecommunication access besides the fees that they charge.

Considering that we already underwrite the telco's with monopoly status, it would be detrimental to allow them to charge websites for access to the customers. Net Neutrality is the only option we can have for situations where monopoly status is given. If the Congress votes to remove monopoly powers of the ISP's as it did with long distance calling than Net Neutrality becomes a moot point as there is now the potential for competition. By allowing, the Inbound ISP's to impose favorable status to some websites and omitting others from access increases their profit margin while they no longer need to invest in the infrastructure. My company ClearVoyager has designed a product that can conceivably allow people to start businesses with state of the art computing structures at roughly 15% of the normal costs, however without Net Neutrality companies like Verizon and TimeWarner can steal our research and development simply by blocking us and charge people alot more than we will for an inferior service. However, until the issue of NetNutrality is resolved once and for all investors are skittish. Who would want to invest in a small company when the mega corporations can simply put your web address in a blocking filter and direct potential customers from your site to theirs or anyone who pays them.

Congress must prohibit the individual states from stopping free wireless broadband, allowing both private companies and municipalities the option to provide this service. Prohibiting the individual states from restricting access does in fact fall under the Commerce clause of the Constitution, as the probability of crossing state lines is high. Allowing private companies to provide wireless access is the option that could have the best and most immediate impact in providing communities and the poor with Internet access. Companies can trademark their networks for advertisement, allowing them to advertise on the login screen could be the incentive to get this private investment going. Having free internet access will only help the poor. Providing free wireless access to school children can help to lower the tax burden on property owners.

While we may dislike some messages on the internet, as Americans we cherish the freedom to say and read things that others may find offensive. Some people and organizations have squashed free speech by threatening the outbound ISP's who hosts these websites with lawsuits. Congress needs to pass a law that prevents individuals or corporations from suing the ISP to shutdown the website unless they have a court order to do so. If there is libelous or defamatory material on these sites the proper recourse is to seek relief from the website's owner. Only if that owner or his agent cannot be served with legal papers should the ISP shutdown that website. Basically the Internet is a town square and outbound ISP's are the soapbox, it is the websites that are the speakers. If someone was to stand on the Mall in Washington and say John Kerry hates Heinz Ketchup and loves Hursts would you sue the federal government for providing a place for that person to speak? Of course not, you would seek relief from the person who said that in a court. It is the same, because someone may not like the message, the ISP should not suffer for providing a venue for those views. You would not go running to the Capitol police to escort that person off the Mall until you had your day in court and proved them to be liars. It is not up to the Capital police to determine what is true or false, same is true with outbound ISP's. Only when the website is found by a court of competent jurisdiction should the ISP drop that website as a customer.

Finally, web sites must be protected from companies like Google that alter the website to allow them to sell their advertisements on it. The crux of the problem with companies like Google that they can take another's work and use it to sell advertisements to competitors. Google says that their customers want this, but they have failed to ask the websites' owners. It is not harmless in what Google is doing, people devote both time and money to present their information in the manner they want, but Google is altering the style and structure of the website. I will give you an example that you can relate to. Going back to 2006, if I searched Google for John Kerry I would find websites for both your supporters and detractors, and this is what it should be, but if I was on to your website, I should reasonably expect to find your message unadulterated. But with Google the term swift boat would be tagged by them and I could see on your website advertisements for the Swift boat veterans for truth. Now I ask you is that fair or has your investment been compromised?

So to sum up my thoughts I would hope to see the following as a comprehensive bill to advance the internet.

1. Limited monopoly on ISP's like Verizon, Cox Cable, TimeWarner, Comcast, etc.
2. Ability for companies and municipalities to act as an ISP provide free wireless access
3. Net Neutrality
4. Protection of outbound ISP's from being a party to the initial lawsuits.
5. Protection of websites from alteration in both content and style by a third party.

Thank you for this opportunity to air my thoughts. Respectfully yours,

Theodore Moran, Partner and Chief Technology Officer
the ClearVoyager Corporation

"Visionaries of the design and delivery of efficient and adaptable virtual application development and testing platforms."

US Office
167 Fort de France Avenue
Toms River, NJ 08757
908-420-0552

UK Office
9 Huckleberry Close
Barton Hills
Luton LU3 4AN
44 1582 510 345

55. _____ cameraone1 Says:
September 20th, 2007 at 11:34 pm

In a lot of countries that I have been to have free Wifi in parks and shopping centers. Even in Taipei there is a blanket of wifi for 7 miles throughout the downtown metro center and was growing. Imagine that free internet in NY City ALOT of people would get connected. I know the US has some of that but usually it is to encourage you to go to there shop but on a more powerful level would be fantastic.

56. Jared Says:
September 20th, 2007 at 11:41 pm

It is critical that all Americans have access to high speed internet. My daughter cannot afford to supply her home with internet service and she lives in adjusted income housing. Her daughters do not have the opportunity to do the research they need for school projects, she cannot take advantage of employment and education opportunities available. There were many promises made, but there has been no pressure from government to make the Internet available for everyone. It only helps to further divide us along economic lines. In cities where local government has tried to offer broadband service, phone and cable companies have spent $$$$$$ to stop them! Tells me, there's a lot of money being made. I believe internet access should be made available like electricity...

57. sandynn43 Says:
September 19th, 2007 at 11:43 pm

I recently caught Hughes.net manipulating my usage to their benefit. I printed the sheets, got an update and 3 free months use. If they "tap" people, they cut their usage down to almost nothing. They do this to enough people, they can sell more space for people to join. Make more money.

At one point they gave me so much satellite space to use, the way I want to use it. Now I can't, I got tapped... They manipulate people.

58. *DaResdoc Says:*
September 29th, 2007 at 15:54 pm

As I understand, efforts by cities like Philadelphia to create a municipal utility to provide free or low-cost broadband wireless to all their residents were made illegal in 2005 by Congress, at the urging of the usual suspects, like Verizon, ATT and Time-Warner.

Why can't cities, towns and counties provide this service and reduce the inflated costs provided by private carriers? The spectrum belongs to us, and private companies should not have the right to monopolize it to themselves.

Other countries provide widespread cheap broadband access. Are we going to keep falling further behind in this technological area too?

59. *chrbrwit Says:*
September 29th, 2007 at 11:54 pm

When did we start hearing about the problem of Internet neutrality? about the time of the ATT/SBC/Ameritech and GTE/Bell Atlantic mergers, which was also about the time that the FCC decided that broadband modems (that's right, the modems themselves) were no longer part of a common carrier telephone system.

It all comes down to the wires. The Internet is not a cloud, it is data moving on wires (ok, there's some wireless in there, but that's basically a last-mile technology), and most of those wires are, or were, built with ratepayer money as part of the public switched telephone network. The telcos, with their armies of lawyers who pour out of the elevators every day at every state and federal regulatory agency across the country, have wrested access to these wires from the common carrier system, and deposited it instead in the virtually unregulated category of "information service." That's the trick. And with that, the way is free for the carriers to create "bottlenecks," exact premium rates for what was normal carriage, and create a network of affiliated vendors who are all tithing to the mother church.

We need to return to common carrier principles, a network that's "dumb in the middle." This is how the Internet became what it is today.

60. *anst Says:*
September 30th, 2007 at 12:06 am

Dear Senator Kerry:

Thank you! Please continue to help spur competition against big phone and cable companies.
I agree with you that "Restoring America as a Broadband Leader" will benefit the country as a whole.

61. *Aefbmoj Says:*
September 30th, 2007 at 12:16 am

I am not competent to propose any ultimate answers, but I observe that the Internet has completely changed the complexion of daily life, information, and commerce in its present form. For those who have access. This is potentially so fundamental a change in the economic and social structure of society, I can not accept that it might be relegated to private enterprise, nor stratified according to economic influence.

While it was a monopoly, the highly regulated telephone system in the last century did, in effect, make telephone service available to everyone and it became the equivalent to a "right" - no one was left out, and while some areas remained the equivalent of "oatmeal cans with strings" for a while, there was affordable access and it was not income dependent.

I agree with the concept of making internet access somewhat the equivalent of a utility, like water or electricity. While perhaps not completely true now, this is, in whatever form it takes, the infrastructure underlying the economy, information, social contact, and social evolution for the forseeable future. We are looking at 2040. It's going to be pretty amazing. We need to lay the groundwork now.

62. *chickster* Says:
September 29th, 2007 at 11:23 am

Through the years, we have watched a number of media slide downhill. First it was AM radio which degenerated into a vast wasteland. Next it was FM radio which today, with very few exceptions, is well on its way down the same path. Both are profit-driven and hence meaningful or thoughtful content is of little value as it generally will not "turn-a-buck". Neither of these have tumbled downward as fast as television whose content is obviously pointed at the mindless millions who dominate the ratings. After all ratings mean dollars. The printed media is no better as it comes more and more under the control of fewer and fewer individuals. In even our largest cities, many are a one newspaper entity. If you happen to agree with the political bent of the local paper, you are happy and everything you read is gospel. If you don't the local written media is a vaccuum into which no air can be injected.

Please do what you can to keep and encourage a free voice and a free America which is not controllable by huge corporations or by any other power structure which seeks to enforce stifling control. Today the internet is the only frontier these forces have not yet controlled.

63. *bobkay* Says:
September 29th, 2007 at 11:44 pm

At 68 years of age this republican became very unhappy with both the Dems and republicans. After a lot of searching, MoveOn has ultimately become my vehicle for pursuing reform and good governance. Key to the success MoveOn creates is the ability for grassroots organizations to have a voice, organize, raise money and take action through the mass voice of the internet. Hence, congressional protection of netneutrality is a must, making available orphaned frequencies (TV, radio, etc) for internet use and breaking up big-media monopolies are critical.

64. *Steve Herzfeld* Says:
September 28th, 2007 at 1:33 pm

Dear Senator Kerry,

I sincerely hope you are able to make a difference on this issue. The United States is not currently leading the world in any really important area except perhaps military spending and exporting of weapons of destruction.

If we want to be world leaders we will have to get our act together and put an end to the system of government run entirely by moneyed interests who own both the Democratic and Republican parties.

We have to spend on infrastructure so our schools are adequate and our bridges don't fall into the rivers. Our railways are badly in need of proper upkeep as well and yes, the internet in America is not as good as it is in many other countries and yes that is the direct result of letting it increasingly become a profit center for some at the expense of it being a powerful business and educational tool for all.

The issue being discussed here is a symptom only. I wish you well and support your efforts as I supported your candidacy.

Steve

65. *FallenTears Says:*
September 29th, 2007 at 1:25 am

Hello, Sen. Kerry.

I have a brief anecdote. I sent a letter to my representative, Elton Gallegly (R-CA-24), outlining my concerns over this issue, as he has voted against the concept of network neutrality wherever possible in the past. I included sources where necessary to indicate I actually had some inkling of what I was talking about, but more importantly, to get a response out of him, as I'd love to see a general response to many of the points either of us have brought up. Of course, I received a form letter in return, as I was moderately confident I would, but—better yet—the form didn't even get my position right! I guess even Gallegly's assistants don't give enough of a damn to put letters received about the issue in the right pile.

I get the feeling that many republicans like Gallegly don't care about this issue—they brush it off as a purely economic one, allowing them to cache it in the schema of "no regulation, regulation = BAD". This is not merely an economic issue—it's a social one.

Some would go as far as to claim that it's an extension of the polarization of rich and poor, and while I think that might be a bit drastic. However, it certainly seems more and more on-the-mark, with corporations like Verizon lobbying against free, municipal wifi for impoverished inner-city areas. This was apparently the case for inner Philadelphia in 2004—at the time, less than half the neighborhoods there even had internet services available—but then-mayor Ed Rendell threw his telecom buddies a bone by signing into law a bill written almost entirely by Verizon.

They used, as justification, the argument that corporations shouldn't have to compete with the government. This sounds reasonable, but what competition is there when the area's primary ISP doesn't even think it could see profit in laying down lines for poor areas? "We won't give them access, but you can't, either"? Is this selfishness or just hubris?

I have one more example. I used to be a subscriber to Adelphia, which Time-Warner-AOL and Comcast purchased/split the customers of after it was busted for... corruption? Anyway, the two purchasing companies purposely bought up plots of the former company's customers in such a way that they could be sure they would never need to compete in those areas. While I suppose this is perfectly legal, it bothers me that we've reached a point where conglomerates seem to be competing with their customers' ISP options more than they are with one another.

At any rate, I don't think much of congress understands how important this is, and how important it's going to become; it's easier to toe the line on issues that could be perceived as economic. A job is needed, something to make them—mainly the republicans, to be blunt—understand just how abysmal an internet that goes the way of television could become.

66. _____ OGrady Says:
September 30th, 2007 at 1:58 am

I agree with the idea that the Internet should be a utility, but I don't think that goes far enough. The ideal situation would be to have free wireless available in homes and businesses (for those who choose it) and in public parks, buildings, and squares. Besides empowering people, this would be good for small businesses as it would allow anyone passing through town to sit down and look up what services are locally available. (This would be especially true if there could be specific local networks set up along with Internet access.)

Currently, however, the only access for people who can't afford computers is in the libraries, which naturally sets up a conflict over resources, since libraries have so many other responsibilities. I would like to see independent public computing centers set up in a manner similar to, but independent from, libraries.

67. lschandler2000 Says:
September 28th, 2007 at 2:07 am

Senator Kerry,

We all know what the Internet was and has become. Too many businesses seem to have their hands in it to keep it slow and out of reach to most people. By most people, I do refer to the lower income people. These are the grassroots of this country and should be getting their fair share.

The cost for access is unbelievably high compared to the cost of hardware and software which is already in place. A part if this is caused by monopolistic holds on public resources. The telephone poles which cable is attached to (as well as phone lines) are public utilities. Since returning to my home town a year ago, the number of internet/cable providers has gone from three to one. This media giant (TW) is gobbling up everyone, and charging whatever they want. I pay $39 for a medium speed internet connection. Dial-up is the cost of a phone line, but those are disappearing wherever cable is run. Our governor (Ohio) has recently signed legislation to force cable companies to provide access across the state.

There is talk of a new internet. This not only needed, but necessary. The type of functionality I have been seeing in their "wish list" will end most of the spam and phish mongers. At the same time, speeds will increase dramatically. This is probably one of the most important features as we are twentieth in the world in high-speed broadband.

Yes Senator, I support you in bringing change to a system that was not designed to do as much as it does today. I implore you to push our leaders into the future, before it is too late. Please, make the information highway a reality not a side street.

Thank you, one of your supporters.

68. Carrie Says:
September 28th, 2007 at 2:31 am

I would like to see a stop to ISPs practice of throttling bandwidth for important technologies such as bit torrent and other P2P applications. They should also stop lying to their customers when they are asked about such practices. Many small business are built around these technologies, and by attempting to limit the flow of bandwidth, they are essentially hurting entrepreneurs in this country. They are not providing the service that their customers are paying good money for. We need to have a choice of using any ISP, so that we aren't forced to use bad ones that don't provide the service that we all need and pay for. Let's open up the internet!

69. heathergirl Says:
September 28th, 2007 at 2:16 am

I have a small business in rural Washington State on an island with a bridge. All cable stops on the mainland. I have dial-up. It is awful for me. I cannot load my website updates. I have to drive 8 miles to the mainland and use my laptop to update my site at a wireless spot at a local bakery. Therefore, I don't do them as often as I am supposed to. With dial-up, I wait forever to download orders from the net and respond to customer's email. I cannot download photos they send me. I wait and wait and waste so much time. Why can't we have wifi max like they do in some third world countries? I understand their coverage is comparable to cell tower coverage. There are several small businesses like mine locally that waste half of their day waiting for the internet. It is killing our opportunities. My husband cannot telecommute and drives 150 miles per day round trip because we have dial-up. It is too slow for his work servers. I can live fine without cable for TV, but the internet is another story.

70. _Swift2 Says:_
September 26th, 2007 at 4:58 am

Others have mentioned wireless, but I really think that is a key. In the cell phone world, we are locked in to providers and their subsidized or crippled phones.

We must get to the same average broadband speed/cost as the Japanese and Koreans. We must end the stranglehold the private ISPs have on the Internet.

71. _mellcraft19 Says:_
September 26th, 2007 at 6:52 am

Senator Kerry,

You know what's right. Many of the ISPs (Comcast, RoadRunner, Verizon, TimeWarner) have gotten access to our homes via local franchises. They dig up our streets, and they occupy our poles. All fine and dandy. The do - in most cases - provide a valuable service.

But this service is under attack. Not only do we lack any serious competition (in most areas there are only two alternative, Cable Co. and Phone Co.) and prices keep going up, the rate of service (speed) is laughable in an international comparison.

What is not laughable though is the cost. We here in the US pay more - for less - than many other western nations.

We can change this. Every provider needs access to the local customer. Fair enough. Let one or a few companies provide the "connectivity" (physical connection and an IP address) as a Utility (regulated). If not, let other companies rent access on the existing networks. They've gotten to those networks via "subsidies" (franchise agreements that are written so no other cable company can enter into a geographical area), not more than fair that they also share the networks.

Then let other companies provide the services. I like Comcast's "physical" network (reliable) but I have no interest in getting my phone service from them, nor in the e-mail service, or their "home page". There are other much better/cheaper alternatives for that (Vonage, Associated Press, Google). Still, I have to pay for that and Comcast has the ability to block these "other" services and when they do, I really do not have an alternative - other than the Phone Co. which in most cases is just a mirror image of my previous provider. Same service, same pricing, just different color.

We need to allow the competition access to the physical network!

We also need to educate our City Governments on "access issues". When they prepare new areas, or dig up old streets, the Cities should also put in (empty) conduits for "communication services". When Comcast, Verizon, or the likes, come to put in their wires, instead of digging up the street, they should be mandated to "rent" space in existing conduits - conduits they have to share with other providers. By "prepping" our infrastructure (roads/sidewalks, etc) from day one, we not only provide less disruption by preventing digging in new or repaved streets, but we are also making it easier for other providers to reach the consumers (the last mile problem).

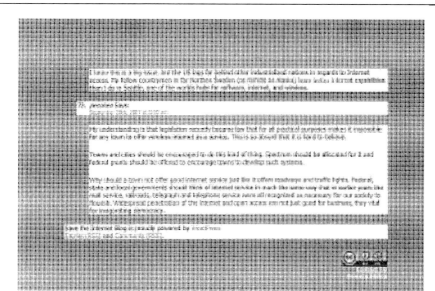

COMMENTS FOR THE RECORD

THE COMPUTING TECHNOLOGY INDUSTRY ASSOCIATION (CompTIA), ROGER, J. COCHETTI, GROUP DIRECTOR, U. S. POLICY

Testimony of

The Computing Technology Industry Association (CompTIA)

Roger J. Cochetti

Group Director-U.S. Public Policy

Before the Senate Committee on Small Business and Entrepreneurship

On

"Improving Internet Access to Help Small Business Compete in a Global Economy"

Wednesday, September 26, 2007

On behalf of CompTIA's more than 20,000 members, I am pleased to submit this testimony for today's hearing, "Improving Internet Access to Help Small Business Compete in a Global Economy." It is our strong belief that the single most important thing this Congress can do to improve small business access to the Internet is to ban taxes on Internet usage or access. As such, we strongly support passage of S.156, the Internet Tax Freedom Act. Our members, particularly our roughly 15,000 small business members, thank the Committee for discussing a topic so vital to the welfare of American small businesses and consumers.

Small businesses are the backbone of the American economy. Some 23 million small businesses employ over half of the private sector workforce. Small businesses are a vital source of the entrepreneurship, creativity, and innovation that keeps our economy globally competitive. As a nation, we are dependent upon the health of the small business sector, and this is why we so adamantly support the moratorium on Internet access and usage taxes and the prohibition on multiple or discriminatory Internet sales and use taxes.

In 1998, the United States Congress championed small businesses and consumers, while promoting growth in the American economy, by enacting a moratorium on Internet access taxes, as well as new, discriminatory taxes on e-Commerce . This was further extended, most recently in 2004. The Congress' wise policy decision in 1998 served as a catalyst to revolutionize the way that American small businesses conduct their business and it should be made permanent to ensure that small businesses remain globally competitive and continue to drive our economy. Nearly all economists today agree that the unprecedented growth in American productivity (with almost no inflation that we experienced from the mid 1990's onward) was driven in large part by the benefits of the use of the Internet by business and consumer alike. These statistics do not begin to capture the additional enormous contributions that the Internet has made to education, culture, entertainment, and international cooperation, however.

The U.S. policy of not permitting anti-Internet taxes—including taxes on Internet access or taxes that discriminate against Internet transactions—has encouraged creative talent, entrepreneurs and investors to develop new businesses and business models on the Internet and it has encouraged small businesses everywhere to make use of the Internet. Today, well over half of all American small businesses rely on the Internet every day for their productivity, marketing, sales, and customer relations. Moreover, the United States has led the world by prohibiting taxes on Internet access and usage and on anti-Internet taxes. The Congress should not be in a position in which this vitally-important law expires in a few weeks and it should not ever be in that position again.

CompTIA Overview.

CompTIA is the largest computer industry trade association in the United States. We include among our members virtually every brand name and large company in the industry as well, as noted above, roughly 15,000 small information technology (IT) companies that are commonly called Value-Added Resellers or VARs. A typical small business in the United States, almost regardless of its business, will not have an IT department of its own. Instead, America's small businesses rely on the services of one or more of thousands of VARs that are located in every city, town, country and district in the United States.   VARs are small businesses who are themselves system integrators and operators. VARs design, install and maintain computer systems and networks for other small businesses. An estimated 32,000 VARs, most of which are small businesses themselves, sell approximately $43 billion dollars worth of computer hardware, software, and services annually. This means that over one third of the computer hardware sold in the U.S. today is sold by VARs.

While we in CompTIA represent all segments of the IT industry, including large hardware, software, services and training companies, we also uniquely represent America's VARs. For 25 years, CompTIA has provided research, networking, and partnering opportunities to its 20,000 mostly American member companies, nearly 75% of which is comprised of American VARs – the small business component of the tech industry

In addition to representing the interests of VARs, CompTIA also works to provide global policy leadership for the IT industry through our headquarters in Chicago and our public policy offices in Washington, Brussels, Hong Kong, and Sao Paulo. For most people in the computer industry, however, CompTIA is well known for the non-policy-related services that it provides to advance industry growth: standards, professional certifications, industry education, and business solutions.

The Online Economy, Small Business and Consumers.

The clients of CompTIA's 15,000 VARs are traditionally not large corporations; our VAR members serve as the IT departments for America's small businesses, which are themselves the backbone of the American economy. As such, our members are highly sensitive to the needs of both small businesses and consumers of small business products and services. At the time of the initial passage of the Internet Tax Freedom Act, many small businesses considered it a relatively expensive novelty to maintain some presence on the Internet. Now, the times have drastically changed.

On May 21st, the new head of the Small Business Administration (SBA), Steve Preston, told a conference of small businesses that embracing high-tech solutions "can mean the difference between maintaining a competitive organization and potentially not being in business anymore." He also explained, as we all know, that something as simple as having a high-quality Web site can "make a small business look like a big business." Small businesses are as much, if not more, a beneficiary of the benefits of the Internet as are large businesses and consumers.

Naturally, in today's U.S markets, small businesses justifiably consider marketing and selling on the web essential to the success of their businesses and are pushing the envelope of innovation and creativity online. In a recent survey of small business owners conducted by allbusiness.com, 83% reported that the Internet had improved communication about their company, and 61% said that the Internet had helped open new markets for their businesses. The same survey found that small businesses are using the Internet to improve operational efficiency.   87% use the Internet for business communications and 89% of those surveyed use the Internet regularly for research. Moreover, 44% of small businesses surveyed already had a website up and running and 38% had plans to launch a website within six months. The Internet is the great leveling field in American business today, permitting a small business to compete with one that is much larger.   In that respect, the Internet promotes competition, innovation and productivity.

We should not discourage the use of this medium by taxing it or allowing anyone to impose discriminatory taxes that will discourage its use. In fact, we should be doing everything that we can to encourage greater access to the Internet and investments in, and use of, it.

Small Business Success Online Depends on Access, Affordability, and Parity.

These successes for small businesses and consumers would not have been possible without the Internet Tax Moratorium. There are two driving factors behind the growth of eCommerce for small business: Access to affordable high speed Internet connections for businesses and consumers and tax parity between online and offline sales. Onerous and unjust Internet tax schemes would add significantly to the cost of both providing and obtaining Internet services and thereby discourages Internet usage and broadband adoption. Should this occur, small businesses and consumers would experience more limited availability of Internet infrastructure, frustrating the Internet's rich promise. For these reasons, CompTIA supports broadband deployment, broadband competition, and the further closing of the digital divide, in addition to supporting a ban on Internet access taxes.

But access must be more than available—it must also be affordable *and predictably affordable*. For small businesses operating on slim margins and consumers working to make ends meet, even small increases in cost can push either group offline. Further, the unpredictability of what new taxes may be imposed on Internet services at any time will scare away consumers, investors and entrepreneurs alike. Layering an unpredictable array of changing Internet access taxes on top of what is an essential but moderately priced component of people's businesses and lives can easily prevent those with the most to gain from the empowerment of the Internet from being able to use it. This is especially true for rural small businesses. According to an SBA study in December of 2005, rural small businesses pay nearly 10% more for broadband services than their urban counterparts. Given that broadband services are price elastic, disparities created by access taxes will unjustly harm small businesses, particularly those in rural areas.

The driving force that the Internet has become for the United States' economy and culture should not ever be subject to an access tax. Such a regressive tax would place a significant hurdle to clear for small to access to what has become the defining economic, political and cultural necessity of this century. Permitting unpredictable and multiple taxes on Internet access was not a sound social or economic policy in 1998, and it will not be in 2008, 2018, or 2028. Congress should provide cost conscious small businesses and consumers the piece of mind that their increasing investments in and reliance upon, the Internet will not be wasted because arbitrary state or local taxes levied against them in the future will make the Internet unaffordable.

Recommendations and Conclusion.

To continue Congressional support of small business, CompTIA specifically supports the following measures:

- The Internet Tax Moratorium should be made permanent or extended for the long term for both access taxes and for new, multiple, unpredictable discriminatory sales and use taxes.

- The Internet Tax Moratorium legislation before the Senate should be amended to clarify the definition of Internet access to cover all the services intended by Congress in enacting the original moratorium and in subsequent amendments. This is necessary to prevent taxing bodies from finding creative ways to try to tax Internet access services.

- The grandfathering of certain states' authority to raise Internet access taxes under the 1998 moratorium should be ended. These states have had sufficient time to identify additional revenue streams and to decrease their dependence on any Internet access taxes they may have been assessing. Just as there should never be a tax to walk into a library, shopping mall or government office, there should never be a tax for accessing information, products, or national, state or local eGovernment offices online, regardless of a businesses operating state.

Given the growth of the Internet's economic and social importance—from saving time buying back to school clothes, to finding and evaluating a doctor or searching for employment—access to the Internet, free of unjust taxes, is one of the most critical issues before America's consumers and small business entrepreneurs. As such, CompTIA strongly encourages this Committee and this Congress to continue its vigorous defense of small businesses and American consumers by passing S. 156.

STATEMENT FOR THE RECORD
BY
WILLIAM MICHAEL CUNNINGHAM
AND
CREATIVE INVESTMENT RESEARCH, INC., WASHINGTON, DC

Creative Investment Research, Inc.
PO Box 15385
Washington, DC 20003-0385
866-867-3795 phone/fax
http://www.minorityfinance.com
www.minoritybank.com
http://www.creativeinvest.com

STATEMENT FOR THE RECORD
by
WILLIAM MICHAEL CUNNINGHAM
and
CREATIVE INVESTMENT RESEARCH, INC.
Submitted to the
U.S. Senate Small Business Committee Field Hearing in
Prince George's County

William Michael Cunningham and Creative Investment Research, Inc. (CIR) submit the following statement for the record of the *U.S. Senate Small Business Committee Field Hearing in Prince George's County.*

We thank U.S. Senator Benjamin L. Cardin for this opportunity and for investigating "the problems that small and minority businesses encounter when attempting to contract with the federal government." We urge the Committee to continue to get opinions on this matter from a culturally and economically diverse set of persons.

We support the Committee's efforts to modernize policies and procedures concerning minority business contracting. We believe the hearing is a proper first step.

Background

William Michael Cunningham registered with the U.S. Securities and Exchange Commission as an Investment Advisor on February 2, 1990. He registered with the D.C. Public Service Commission as an Investment Advisor on January 28, 1994. Mr. Cunningham manages an investment advisory and research firm, Creative Investment Research, Inc.

Creative Investment Research, Incorporated, a Delaware corporation, was founded in 1989 to expand the capacity of capital markets to provide capital, credit and financial services in minority and underserved areas and markets. We have done so by creating new financial instruments and by applying existing financial market technology to underserved areas. The Community Development Financial Institution Fund of the US Department of the Treasury certified the firm as a Community Development Entity on August 29, 2003. The Small Business Administration certified the firm as an 8(a) program participant on October 19, 2005. We have not received any revenue due to our participation in the 8(a) program.

In 1991, Mr. Cunningham created the first systematic bank analysis system using social and financial data, the Fully Adjusted Return® methodology. In 1992, he developed the first CRA securitization, a Fannie Mae MBS security backed by home mortgage loans originated by minority banks and thrifts. .

In 2001, he helped create the first predatory lending remediation/repair MBS security.[1]

Mr. Cunningham also served as Director of Investor Relations for a New York Stock Exchange-traded firm. On November 16, 1995, his firm launched one of the first investment advisor websites. He is a member of the CFA Institute

Pool	Client	Originator	Social Characteristics
FN374870	Faith-based Pension Fund	National Mortgage Broker	Mortgages originated by minority and women-owned financial institutions serving areas of high social need.
FN296479			
FN300249			
GN440280	Utility Company Pension Fund		
FN374869		Minority-owned financial institutions	
FN376162			
FN254066	Faith-based Pension Fund	Local bank	Predatory lending remediation

and of the Twin Cities Society of Security Analysts, Inc.

The firm and Mr. Cunningham have long been concerned with the integrity of the securities markets. We note the following:

- On Monday, April 11, 2005, Mr. Cunningham spoke on behalf of investors at a fairness hearing regarding the $1.4 billion dollar Global Research Analyst Settlement. The hearing was held in Courtroom 11D of the Daniel Patrick Moynihan United States Courthouse, 500 Pearl Street, New York, New York. _No other investment advisor testified at the hearing._

The firm and Mr. Cunningham have long been familiar with "the problems that small and minority businesses encounter when attempting to contract with the federal government." We note a few of our experiences below:

- On 6/15/98, the Government National Mortgage Association (GNMA), part of the Department of Housing and Urban Development, issued RFP GNMA 98-PP-02. The RFP solicited various business advisory services, market research, issuer training sessions, job performance enhancement sessions on industry issues, and survey development and analysis. The RFP indicated that the bids would be evaluated awarded in accordance with FAR contracting rules and regulations. Creative Investment Research, Inc. was notified on 8/12/98 that we were an "unsuccessful offeror under the subject solicitation." We were further notified that "While an award has not yet been made, your firm has been eliminated from any further consideration for award based upon a comprehensive review and analysis of all proposals received." In short, we would not be allowed to bid on this contract, although we complied fully with published RFP selection criteria. The contract award was motivated by factors not indicated in the RFP, evaluation factors that changed after the RFP was issued.

- The U.S. Department of Transportation issued RFQ DTTS 59-98-Q-00011 on June 17, 1998 requesting a contractor to:

1) Provide an independent analysis to OSDBU in the review of the financial condition/performance of the commercial banks participating in the STLP;2) Develop criteria for use in the selection of additional lead/participating banks if the program is expanded or replacement banks are required;3) Provide independent banking/loan review of STLP recommendations provided by the participating banks;4) Participate in the program review of the OSDBU financial assistance programs; 5) Consult and provide advise to the Director, OSDBU.

The RFQ indicated that the bids would be evaluated awarded in accordance with FAR contracting rules and regulations. The contract was awarded on 6/23/98 to another firm using undisclosed contract award criteria.

- On October 7, 2005, the House Financial Services Committee requested that the Government Accountability Office (GAO) "examine the federal banking agencies' current efforts to promote and preserve minority-owned financial institutions and the views of the minority financial services community on the effectiveness of these efforts." This involves reviewing federal banking agencies' implementation of section 308 of the Financial Institutions Reform, Recovery, and Enforcement Act of 1989 (FIRREA). As an 8(a) firm, on December 14, 2005, we submitted a proposal to the General Accountability Office to assist the agency in the completion of this study. We have unique and detailed ratings and information on minority banks dating back to 1991. GAO replied that the agency did not wish to contract with an outside firm concerning this matter. Less than a month later, the Agency contracted with a non minority professional to obtain, at greater cost, the information and services we offered to provide.

- Rather than support and engage in the types of predatory subprime lending practices that have negatively impacted the mortgage market and the country as a whole, we proposed to develop alternative, socially responsible methods to enhance homeownership opportunities for minorities and women. As an

8(a) firm, we submitted an unsolicited proposal to Department of Housing and Urban Development (HUD) on April 7, 2006. In our proposal, we offered to research and create a collaborative, market-based approach to increase market participation in a HUD-based socially responsible mortgage lending program. HUD replied that the "Office of Policy Development and Research (to whom we submitted the proposal) is not in a position to support this activity."

It is our belief that federal government contracting and capital market practices, in general, are deeply flawed. It is our hope that the Committee will begin to review market practices from a systemic, global perspective, since defective practices in one sector have been linked to faulty practices in other capital market sectors:

- In multiple cases, corporate management used fraud and deceptive practices to unfairly transferred value from outsider to insider shareholders.

- Investment analysts issue biased research reports to curry favor with management.

- Rating agencies issue defective research reports. These institutions are supposed to "base their ratings largely on statistical calculations of a borrower's likelihood of default," but one news report noted that:

  "Dozens of current and former rating officials, financial advisers and Wall Street traders and investors interviewed by The Washington Post say the (NRSRO) rating system has proved vulnerable to subjective judgment, manipulation and pressure from borrowers. They say the big three are so dominant they can keep their rating processes secret, force clients to pay higher fees and fend off complaints about their mistakes."[2]

- Pension consultants are, also, conflicted and compromised. "Many pension plans rely heavily on the expertise and guidance

---

[2] "Borrowers Find System Open to Conflicts, Manipulation" by Alec Klein, The Washington Post, Monday, November 22, 2004; Page A1.

of pension consultants in helping them to manage pension plan assets," but, according to an SEC report[3],

> "Concerns exist that pension consultants may steer clients to hire certain money managers and other vendors based on the pension consultant's (or an affiliate's) other business relationships and receipt of fees from these firms, rather than because the money manager is best-suited to the clients' needs."

Envy, hatred, and greed continue to flourish in certain capital market institutions, propelling ethical standards of behavior downward. Statistical models created by the firm show the probability of system-wide market failure has increased over the past eight years. Without meaningful reform there is a small, but significant and growing, risk that our economic system will simply cease functioning.[4]

Fully identifiable entities engaged in illegal activities. They have, for the most part, evaded prosecution of any consequence. We note that the Goldman Sachs, fined $159.3 million by the U.S. Securities and Exchange Commission for various efforts to defraud investors, subsequently received $75 million in Federal Government tax credits.[5]

We also note that Alliance Capital Management, fined $250 million by the Commission for defrauding mutual fund investors, received a contract[6] in August, 2004 from the U.S Department of the Interior (DOI) Office of the Special Trustee for American Indians, to manage $404 million in Federal Government trust funds.[7]

---

[3] *Staff Report Concerning Examinations of Select Pension Consultants*. The Office of Compliance Inspections and Examinations, U.S. Securities and Exchange Commission. May 16, 2005.

[4] Proportional hazard models created by the firm and reflecting the probability of system wide market failure first spiked in September, 1998. The models spiked again in January and August, 2001. They have continued, in general, to trend upward, indicating a heightened risk of catastrophic market failure due to corporate fraud and malfeasance.

[5] The tax credits were awarded under the U.S. Department of the Treasury New Markets Tax Credit (NMTC) Program. (See: http://www.cdfifund.gov/programs/nmtc/).

[6] Contract number NBCTC040039.

[7] The contract was awarded despite the fact that placing Alliance Capital Management in a position of trust is, given the Commission's enforcement action, inconsistent with common sense, with the interests of justice and efficiency and with the interests of Indian beneficiaries. Alliance is also in violation of DOI Contractor Personnel Security & Suitability Requirements.

Recently, we have observed several cases where corporate management unfairly transferred value from outsider to insider shareholders.[8] These abuses have been linked to the abandonment of ethical principles noted earlier. Faulty market practices mask a company's true value and misallocate capital by moving investment dollars from deserving companies to unworthy companies.

Together these practices threaten the integrity of securities markets. Individuals and market institutions with the power to safeguard the system, including investment analysts and rating agencies, have been compromised. Few efficient, effective and just safeguards are in place.

Investors and the public are at risk.

We understand that, given any proposed legislation, crimes will continue to be committed.[9] These facts lead some to suggest that regulatory authorities may have been "captured" by the entities they regulate.[10] We note that under the "regulatory capture" market structure regime, the public interest is not protected.

We favor efforts to increase fairness in our capital markets while opposing

---

[8] Including, but not limited to, Adlephia Communications, the aforementioned Alliance Capital Management, American Express Financial, American Funds, AXA Advisors, Bank of America's Nations Funds, Bank One, Canadian Imperial Bank of Commerce, Canary Capital, Charles Schwab, Cresap, Inc., Empire Financial Holdings, Enron, Federated Investors, FleetBoston, Franklin Templeton, Fred Alger Management, Freemont Investment Advisors, Gateway, Inc., Global Crossing, H.D. Vest Investment Securities, Heartland Advisors, Homestore, Inc., ImClone, Interactive Data Corp., Invesco Funds Group Inc., Janus Capital Group Inc., Legg Mason, Limsco Private Ledger, Massachusetts Financial Services Co., Millennium Partners, Mutuals.com, PBHG Funds, Pilgrim Baxter, PIMCO, Prudential Securities, Putnam Investment Management LLC, Raymond James Financial, Samaritan Asset Management, Security Trust Company, N.A., State Street Research, Strong Mutual Funds, Tyco, UBS AG, Verus Investment Partners, Wachovia Corp., and WorldCom. Accounting firms, including Arthur Andersen and Ernst & Young aided and abetted efforts to do so. We believe there are hundreds of other cases.

[9] We assume that "employees are 'rational cheaters,' who anticipate the consequences of their actions and (engage in illegal behavior) when the marginal benefits exceed costs." See Nagin, Daniel, James Rebitzer, Seth Sanders and Lowell Taylor, "Monitoring, Motivation, and Management: The Determinants of Opportunistic Behavior in a Field Experiment, *The American Economic Review*, vol. 92 (September, 2002), pp 850-873.

[10] See George J. Stigler, "The Theory of Economic Regulation," in *The Bell Journal of Economics and Management Science*, vol. II (Spring 1971), pp. 3-21.

reform for reform's sake.

We cite the following:

"Falsification and fraud are highly destructive to free-market capitalism and, more broadly, to the underpinnings of our society. Above all, we must bear in mind that the critical issue should be how to strengthen the legal base of free market capitalism: the property rights of shareholders and other owners of capital. Fraud and deception are thefts of property. In my judgment, more generally, unless the laws governing how markets and corporations function are perceived as fair, our economic system cannot achieve its full potential. "

> Testimony of Mr. Alan Greenspan, Chairman of the Federal Reserve Board, Federal Reserve Board's semiannual monetary policy report to the Congress. Before the Committee on Banking, Housing, and Urban Affairs, U.S. Senate. July 16, 2002.

We agree.

Summary Comments

The hearing will give minority businesses an opportunity to meaningfully comment on "the problems that small and minority businesses encounter when attempting to contract with the federal government." Capital is the issue of highest importance to minority businesses. Below, we outline a strategy to move capital into minority businesses.

Microcredit Stock Exchange

On Tuesday, October 17, 2006, Chicago's two major futures exchanges, the Chicago Mercantile Exchange and the Chicago Board of Trade, announced an $8 billion merger. The NYSE and other exchanges are in the throes of mergers. We think the U.S. Congress should impose specific community development goals on stock and commodity exchanges, much like those that the Community Reinvestment Act (CRA) imposes on banks. (When banks merge, CRA regulations require banking authorities to certify that the banks involved do not have a history of discriminating against persons of color or low income persons.)

CRA has stimulated billions of dollars of profitable, high social impact lending, provided to underserved communities nationwide. In this way, the

Act encourages depository institutions (banks and thrifts) to help meet the credit needs of the communities in which they operate.

Given this, exchanges should be required to help meet the capital needs of small, disadvantaged businesses. To do this, we suggest the Congress mandate the creation of a Minority Business Micro Stock Exchange, modeled on the work pioneered by the 2006 winner of the Nobel Peace Prize, Muhammad Yunus and the Grameen Bank. Equity capital, or shares in very small (micro) minority businesses would be traded on a Micro Business Stock Exchange.

The Exchange would provide the framework for the provision of small amounts of equity capital. To make things easier and to enhance the probability of success, we suggest the initiative focus specifically on disadvantaged businesses operating in Prince Georges County.

The mechanics are simple: small businesses with capital needs prepare business ready financing proposals that are put before investors on a trading floor managed by the Exchange. Investors review the businesses and their plans and decide whether or not to invest. The plans and the businesses themselves would be authenticated by a set of independent third parties, say, the County Treasurers' Office, and representatives from the local Minority Business Opportunity Center. Terms of any investment would be determined by a standardized micro business investment contract, much like a small business futures contract. The contract would allow for off exchange, "on the curb" modification and tailoring.

Our suggestions are specific and fit well within the business activities and framework of the exchanges.

This is just a very rough idea, in need of refinement. Perhaps the free market economists at my alma mater, the University of Chicago, could be persuaded to help. After all, this Exchange puts free market theories to their ultimate test: if legitimate, legal free market institutions don't work in Prince Georges County, why would they work in, say, Iraq?

In summary, we believe the use of new capital access tools will significantly reduce costs and increase the flow of capital to all sectors in society. This increase in capital access will, in turn, result in significantly increased general economic activity. We estimate, using proprietary economic models, this increased economic activity at $6 trillion dollars over ten years. (This assumes an internet based capital access system that is gender and racially neutral, operating without significant falsification and fraud.)

The internet is a powerful tool. We understand both the potential benefits and the potentially disruptive nature of this technology better than most.[11]

Capital market regulators in other regions of the world will, at some point, enhance their ability to access capital using internet-based tools. Thus, competitive advantage with respect to capital access is available to any country with significant economic potential and a modest communications infrastructure.

We do not know which countries will be winners over the long term. We know with certainty, however, that unless small and minority businesses encounter fewer problems when attempting to contract with the federal government, given the corporate fraud and malfeasance cited, it is unlikely that the United States will long maintain and enjoy its current advantage. The hearings are an important first step.

We look forward to reviewing the Committee's continuing efforts to carry out its mission. We appreciate the time and effort the Committee and the Chairman have devoted to this task. Thank you for your leadership.

---

[11] Our first website, www.ari.net/cirm, went live on November 16, 1995. We appreciate the nature of the task facing legislators. Implementing the proposed modification is very much like performing surgery on a marathon runner - during a race. Corporate fraud and malfeasance threaten the entire system, just as cholesterol clogged arteries threaten the health of the aforementioned runner. To make matters worse, (and to extend this analogy far too long) the nature of the technology is such that it significantly improves the performance of every runner in the race.

INDEX

D

E

F

U

V

W

Y